It's another great book from CGP...

OK, so GCSE Maths can be seriously challenging — and the latest exams are tougher than ever. The only way to master the skills you'll need is to practise answering questions until you're totally confident.

Luckily, this fantastic CGP Workbook is bursting with hundreds of practice questions covering every topic — all completely up-to-date for the new GCSE requirements.

You can thank us later. For now, we'd recommend getting stuck in...

CGP — still the best! ☺

Our sole aim here at CGP is to produce the highest quality books — carefully written, immaculately presented and dangerously close to being funny.

Then we work our socks off to get them out to you — at the cheapest possible prices.

Contents

Section One — Number

Section Two — Algebra

Section Three — Graphs

Throughout the book, the more challenging questions are marked like this: **Q1**

Section Four — Ratio, Proportion and Rates of Change

Section Five — Shapes and Area

Section Six — Angles and Geometry

Section Seven — Probability and Statistics

Published by CGP

Illustrated by Ruso Bradley, Lex Ward and Ashley Tyson

From original material by Richard Parsons

Contributors:
Gill Allen,
Margaret Carr,
Barbara Coleman,
JE Dodds,
Mark Haslam,
John Lyons,
Gordon Rutter,
Dave Williams

Updated by: Joanna Daniels, Emily Garrett, Andy Park, David Ryan, Ruth Wilbourne

With thanks to Paul Jordin and Samantha Walker for the proofreading.

Printed by Elanders Ltd, Newcastle upon Tyne.
Clipart from Corel®

Types of Number and BODMAS

Q1 Put these numbers in ascending (smallest to biggest) order.

a) 1272 231 817 376 233 46 2319 494 73 1101

....

b) 3.42 4.23 2.43 3.24 2.34 4.32

....

Q2 Write down the value of the number 4 in each of these.

For example: 408 ......400......

a) 347 .... **c)** 64 098 .... **e)** 1.64 ....

b) 6754 .... **d)** 745 320 .... **f)** 53.42 ....

Q3 Write down the following without using a calculator:

a) 4^2 = ... **c)** 12^2 = ... **e)** 3^3 = ...

b) 7^2 = ... **d)** 2^3 = ... **f)** 10^3 = ...

Q4 Circle the numbers which are integers.

5 $6\frac{3}{4}$ 7.802 -87 -0.0003 167 π

Q5 Without using a calculator, work out the following:

You'll need BODMAS for this question.

a) $2 + 3 \times 6$ = ... **d)** $7 + 4^2$ = ...

b) $20 \div 4 + 1$ = ... **e)** $(12 + 8) \div (3 - 1)$ = ...

c) $(8 + 3) \times 3$ = ... **f)** $\sqrt{16} \times (12 + 1)$ = ...

Q6 Without using a calculator, find the reciprocal of $\sqrt{(2+3)\times 20-19}$.

OK, don't panic. Use BODMAS to do the calculation. Then, reciprocal just means 1 ÷ the number.

....

Wordy Real-Life Problems

To answer wordy questions, you need to work out what the question is actually asking for. Underline the important pieces of information — the rest of it is mostly waffle. Mmm waffles....

Q1 Michael goes shopping. At the start of his shopping trip he has £20.76 in his wallet. He buys 6 apples at 32p each, 2 loaves of bread at £1.09 each and a chicken for £9.69. How much money does he have left after his shopping trip?

....

Q2 Neil is booking a 14-night holiday.
Package A charges £79.99 per night.
Package B charges £74.99 per night, plus a £90 booking fee.

Which package would be the cheaper option?

....

Q3 Hayley buys a mobile phone for £150. She then has to pay a fixed monthly charge for 18 months. After the 18 months, Hayley has spent a total of £366 on the mobile phone, including the cost of buying it. How much did she have to pay each month?

....

Q4 A teacher has 4 classes of 31 students. She needs to buy 2 pens for each student she teaches. The pens she would like to buy come in packs of 12 and cost £3.89 per pack. How much would she need to spend to buy enough pens for her students?

....

Q5 Martin runs a market stall that sells T-shirts. He pays £2.50 for each T-shirt. He then sells the T-shirts in packs of 5 for £20 each. Martin wants to make a profit of at least £320 each day. What is the minimum number of packs of T-shirts he needs to sell each day?

....

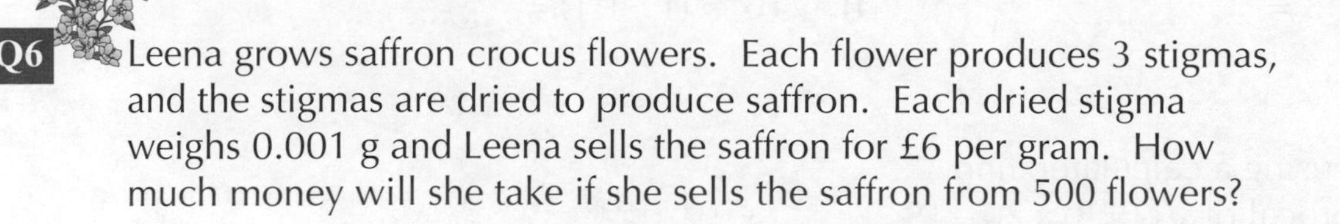

Q6 Leena grows saffron crocus flowers. Each flower produces 3 stigmas, and the stigmas are dried to produce saffron. Each dried stigma weighs 0.001 g and Leena sells the saffron for £6 per gram. How much money will she take if she sells the saffron from 500 flowers?

....

Multiplying and Dividing by 10, 100, etc.

<u>Multiplying</u> by 10, 100 or 1000 moves the decimal point 1, 2 or 3 places to the right — you just fill the rest of the space with zeros. <u>Do not use a calculator</u> for this page.

Fill in the missing numbers.

Q1 **a)** 6 × ☐ = 60 **c)** 0.07 × ☐ = 7

b) 6 × ☐ = 6000 **d)** 0.07 × ☐ = 70

Q2 **a)** 8 × 10 = ... **d)** 0.2 × 10 = ... **g)** 20 × 30 = ...

b) 34 × 100 = ... **e)** 6.9 × 100 = ... **h)** 40 × 700 = ...

c) 436 × 1000 = ... **f)** 4.73 × 1000 = ... **i)** 18 000 × 500 = ...

Q3 A school buys calculators for £2.45 each. How much will 100 cost?

....

Q4 A shop bought 700 bars of chocolate for £0.43 each. How much did they cost altogether?

....

To <u>divide</u> by 10, 100 or 1000, move the decimal point 1, 2 or 3 places to the <u>left</u> and remove zeros after the decimal point.

Q5 **a)** 30 ÷ 10 = ... **d)** 61.5 ÷ 100 = ... **g)** 80 ÷ 20 = ...

b) 5.8 ÷ 10 = ... **e)** 29.6 ÷ 1000 = ... **h)** 480 ÷ 40 = ...

c) 400 ÷ 100 = ... **f)** 753.6 ÷ 1000 = ... **i)** 63.9 ÷ 300 = ...

Q6 Blackpool Tower is 158 m tall. A model of it is being built to a scale of 1 : 100. How tall should the model be?

....

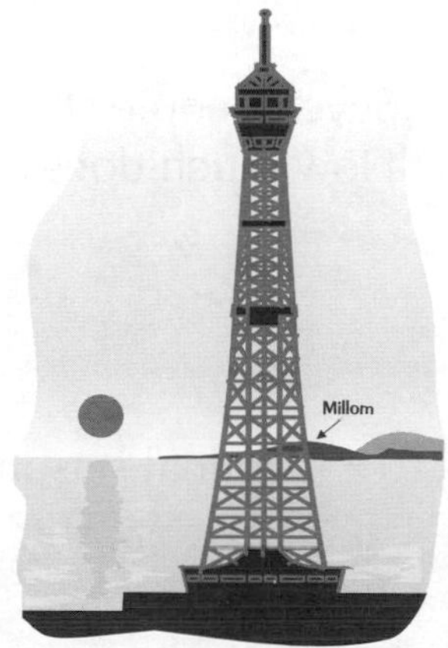

Q7 A box of 40 chocolates contains 56 g of saturated fat. How much saturated fat is there in 1 chocolate?

....

Q8 Mark went on holiday to France. He exchanged £300 for 342 euros to spend while he was there. How many euros did he get for each £1?

....

Multiplying and Dividing Whole Numbers

You're going to have to do a non-calculator exam, so make sure you can do this whole page <u>without using your calculator</u>.

Q1 Multiply the following:

a) 53 × 4 **b)** 89 × 7 **c)** 65 × 38

= .... = ... = ...

Q2 At a petrol station each pump shows a ready reckoner table. Complete the table when the cost of unleaded petrol is 121p per litre.

Litres	1	6	14	28	47
Cost in pence	121				

Q3 Jason can run 318 metres in 2 minutes. How far will he run if he keeps up this pace for:

a) 36 minutes **b)** 54 minutes **c)** 1 hour 12 minutes

....

Q4 Calculate the following divisions:

a) 834 ÷ 3 **b)** 702 ÷ 6 **c)** 286 ÷ 11

If the number doesn't divide exactly into the first digit, you have to carry the remainder over.

Q5 Seven people share the cost of a meal between them. The meal costs £112. How much does each person have to pay?

....

Q6 12 identical satsumas weigh 876 g. How much does 1 satsuma weigh?

....

Q7 Sarah has 2149 screws. She wants to put them into 15 boxes, with the same number of screws in each box. If she puts the maximum possible number in each box, how many screws will she have left over?

....

Multiplying and Dividing with Decimals

Do this page <u>without a calculator</u>.

Q1 Carry out the following multiplications:

a) 3.2×4 **b)** 263×0.2 **c)** 5.7×0.31 **d)** 2.4×3.3

= .... = ... = ... = ...

e) 0.8×6.14 **f)** 2.06×1.4 **g)** 3.27×1.9 **h)** 4.32×2.7

= .... = ... = ... = ...

Q2 Work out the following divisions:

a) $15.6 \div 4$ **b)** $21.6 \div 9$ **c)** $20.8 \div 8$ **d)** $73.5 \div 3$

e) $0.42 \div 0.06$ **f)** $5 \div 0.25$ **g)** $1.56 \div 0.12$ **h)** $2.91 \div 0.006$

Q3 A joiner drills six equally spaced holes in a length of wood.
The centres of the first and last holes are 6.95 m apart.
What is the distance between the centres of each hole and the next?

6.95 m

....

Q4 Steve is ordering some new garden fencing online. He knows what size he wants in yards, feet and inches, but the website gives sizes in centimetres.
Using the information in the table, work out:

a) The number of centimetres in 3 inches.

....

b) The number of centimetres in 6 feet.

....

c) The number of centimetres in 45 yards.

....

<u>Conversions</u>

1 inch = 2.54 centimetres

1 foot = 12 inches

1 yard = 3 feet

1 metre = 3.28 feet

Negative Numbers

Draw out a number line if you're getting confused with the negative numbers. Once you've had a bit of practice, they'll soon become a doddle.

Q1 Write these numbers in the correct position on the number line below:

-4   3   2   -3   -5   1

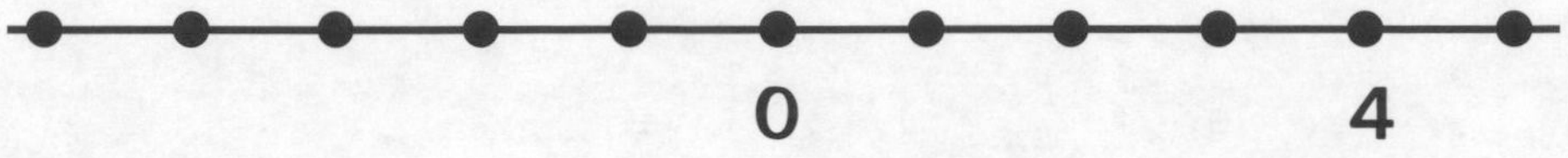

Q2 Put the correct symbol, < or >, between the following pairs of numbers:

a) 4   -8   **b)** -6   -2   **c)** -3.6   -3.7

< means 'less than',
> means 'more than'.

Q3 Rearrange the following numbers in order of size, largest first:

-2   2   0.5   -1.5   -8   ....

Q4 **a)** The weather forecast says that the current temperature is -3 °C, but could reach -5 °C in the next few hours. What would this temperature change be?

....

b) The next day, the forecast says that the temperature is currently -4 °C and will fall by 5 °C overnight. What is the forecast overnight temperature?

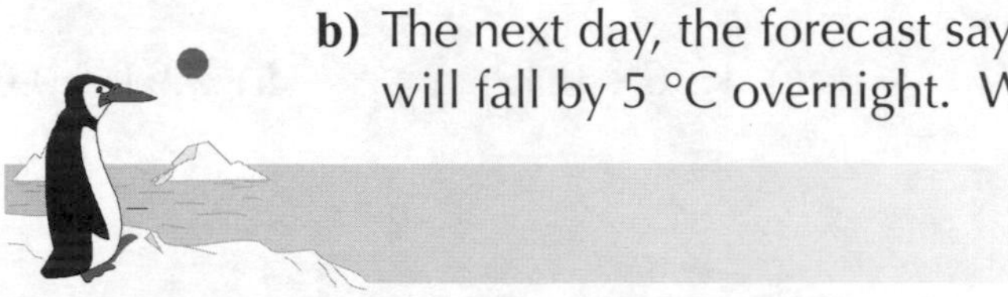

....

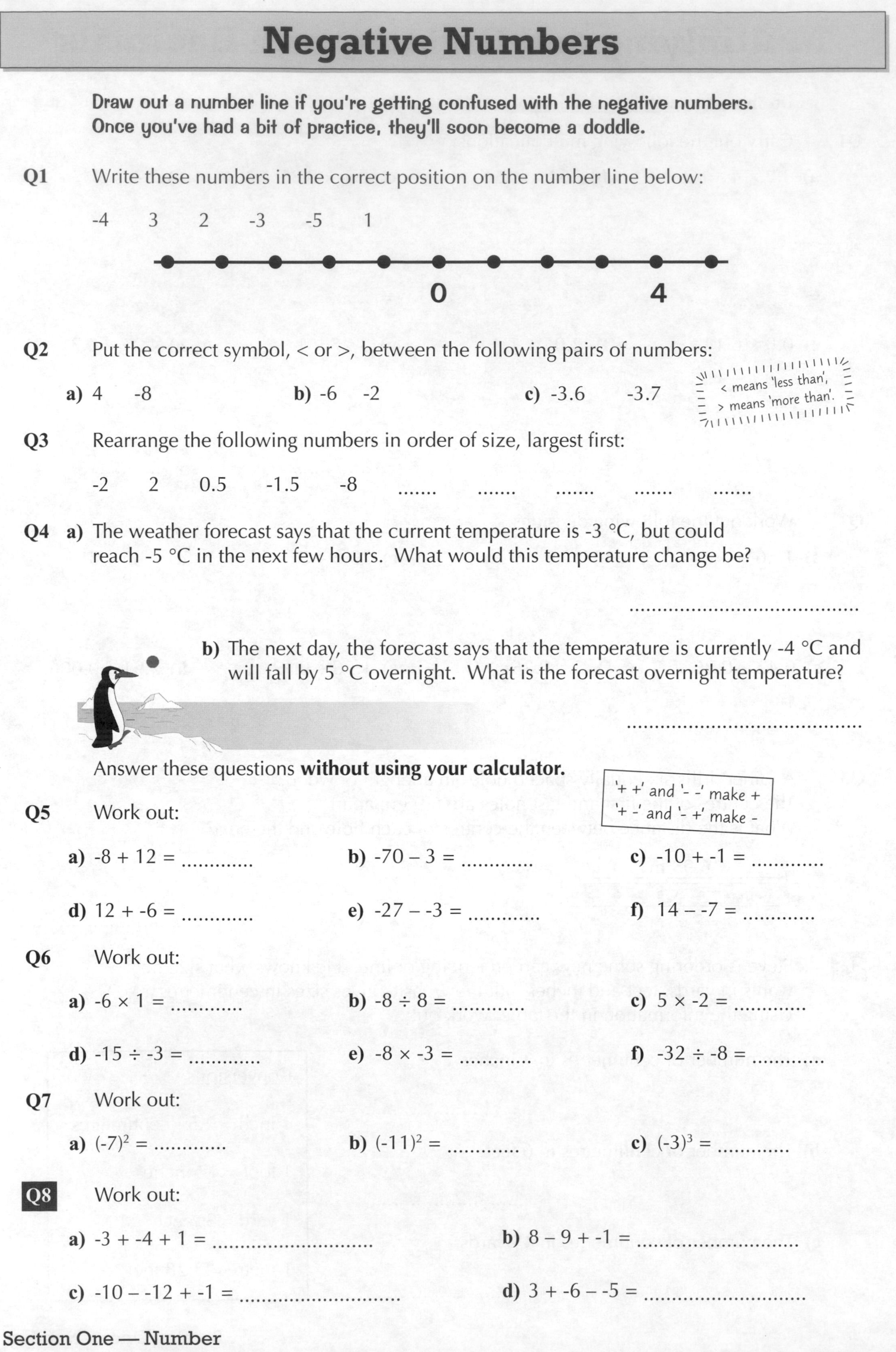

Answer these questions **without using your calculator.**

'+ +' and '- -' make +
'+ -' and '- +' make -

Q5 Work out:

a) -8 + 12 = ...   **b)** -70 – 3 = ...   **c)** -10 + -1 = ...

d) 12 + -6 = ...   **e)** -27 – -3 = ...   **f)** 14 – -7 = ...

Q6 Work out:

a) -6 × 1 = ...   **b)** -8 ÷ 8 = ...   **c)** 5 × -2 = ...

d) -15 ÷ -3 = ...   **e)** -8 × -3 = ...   **f)** -32 ÷ -8 = ...

Q7 Work out:

a) $(-7)^2$ = ...   **b)** $(-11)^2$ = ...   **c)** $(-3)^3$ = ...

Q8 Work out:

a) -3 + -4 + 1 = ...   **b)** 8 – 9 + -1 = ...

c) -10 – -12 + -1 = ...   **d)** 3 + -6 – -5 = ...

Prime Numbers

Basically, prime numbers don't divide by anything (except 1 and themselves).
5 is prime — it'll only divide by 1 or 5. 1 itself is NOT a prime number — remember that!

Q1 Write down the first ten prime numbers. ....

Q2 Give a reason for 27 not being a prime number. ....

Q3 Using any or all of the figures **1, 2, 3, 7** write down:

a) the smallest prime number ....

b) a prime number greater than 20 ....

c) a prime number between 10 and 20 ....

d) two prime numbers whose sum is 20 .... , ...

e) a number that is not prime ....

Q4 Find all the prime numbers between 40 and 50. ....

Q5 In the ten by ten square opposite, ring all the prime numbers.
The first three have been done for you.

1	2	3	4	5	6	7	8	9	10
11	12	13	14	15	16	17	18	19	20
21	22	23	24	25	26	27	28	29	30
31	32	33	34	35	36	37	38	39	40
41	42	43	44	45	46	47	48	49	50
51	52	53	54	55	56	57	58	59	60
61	62	63	64	65	66	67	68	69	70
71	72	73	74	75	76	77	78	79	80
81	82	83	84	85	86	87	88	89	90
91	92	93	94	95	96	97	98	99	100

Q6 A school ran three evening classes: judo, karate and kendo.
The judo class had 29 pupils, the karate class had 27 and the kendo class 23.
In which classes would the teacher have difficulty dividing the pupils into equal groups?

....

Q7 Find three sets of three prime numbers which add up to the following numbers:

10 ...., ..., ... **29** ...., ..., ... **41** ...., ..., ...

Multiples, Factors and Prime Factors

The multiples of a number are its times table — if you need multiples of more than one number, do them separately then pick the ones in both lists.

Q1 What are the first five multiples of:

a) 4? ....

b) 7? ....

c) 12? ....

d) 18? ....

A quick way to do these is just to multiply the numbers together.

Q2 Find a number which is a multiple of:

a) 2 and 6 ....

b) 7 and 5 ....

c) 2 and 3 and 7 ....

d) 4 and 5 and 9 ....

Q3 Steven is making cheese straws for a dinner party. There will be **either** six or eight people at the party. He wants to be able to share the cheese straws out equally. How many cheese straws should he make?

There's more than one right answer to this question.

....

Q4 **a)** Find a number which is a multiple of 3 and 8 ....

b) Find another number which is a multiple of 3 and 8 ....

c) Find another number which is a multiple of 3 and 8 ....

Q5 Which of the numbers 14, 20, 22, 35, 50, 55, 70, 77, 99 are multiples of:

a) 2? .... **b)** 5? ....

c) 7? .... **d)** 11? ....

Multiples, Factors and Prime Factors

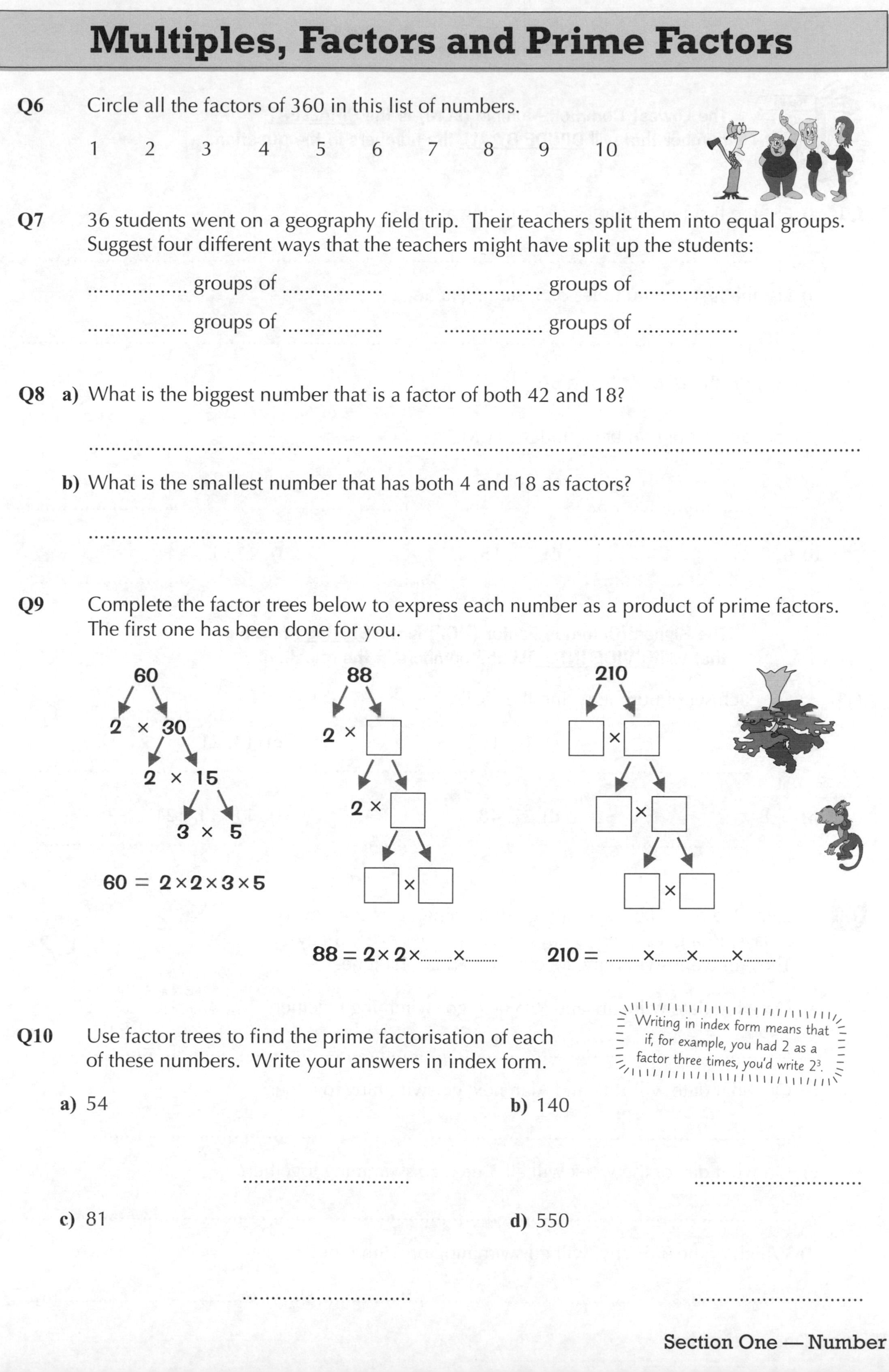

Q6 Circle all the factors of 360 in this list of numbers.

1 2 3 4 5 6 7 8 9 10

Q7 36 students went on a geography field trip. Their teachers split them into equal groups. Suggest four different ways that the teachers might have split up the students:

.... groups of .... groups of ....

.... groups of .... groups of ....

Q8 a) What is the biggest number that is a factor of both 42 and 18?

....

b) What is the smallest number that has both 4 and 18 as factors?

....

Q9 Complete the factor trees below to express each number as a product of prime factors. The first one has been done for you.

88 = 2× 2×........×........ 210 = ...×........×........×........

Q10 Use factor trees to find the prime factorisation of each of these numbers. Write your answers in index form.

Writing in index form means that if, for example, you had 2 as a factor three times, you'd write 2^3.

a) 54

....

b) 140

....

c) 81

....

d) 550

....

LCM and HCF

The Lowest Common Multiple (LCM) is the SMALLEST number that will DIVIDE BY ALL the numbers in the question.

Q1 **a)** List the first ten multiples of 6, starting at 6.

....

b) List the first ten multiples of 5, starting at 5.

....

c) What is the LCM of 5 and 6? ....

Q2 For each set of numbers, find the LCM.

a) 3, 5 ....

b) 6, 8 ....

c) 10, 15 ....

d) 15, 18 ....

e) 14, 21 ....

f) 11, 33, 44 ....

The Highest Common Factor (HCF) is the BIGGEST number that will DIVIDE INTO ALL the numbers in the question.

Q3 For each set of numbers, find the HCF.

a) 3, 5 ....

b) 6, 8 ....

c) 10, 15 ....

d) 27, 48 ....

e) 14, 21 ....

f) 11, 33, 121 ....

Q4 Lars, Rita and Alan regularly go swimming. Lars goes every 2 days, Rita goes every 3 days and Alan goes every 5 days. They all went swimming together on Friday 1st June.

This is an LCM question in disguise.

a) On what date will Lars and Rita next go swimming together?

....

b) On what date will Rita and Alan next go swimming together?

....

c) On what day of the week will all 3 next go swimming together?

....

d) Which of the 3 (if any) will go swimming on 15th June?

....

LCM and HCF

The prime factorisation method can seem a bit daunting — especially if you have to work out the prime factors yourself. But don't panic — you can use a factor tree for that.

Q5 **a)** Write 36 and 24 as products of their prime factors.

36 = ... 24 = ...

b) Use these prime factors to find the LCM of 36 and 24.

....

c) Use these prime factors to find the HCF of 36 and 24.

....

Q6 Use the prime factorisation method to find:

a) the LCM of 16 and 60

....

b) the HCF of 90 and 120

....

c) the LCM of 48 and 72

....

d) the HCF of 28 and 42

....

Q7 Evan is making ham and cheese sandwiches for a party. Each sandwich is made from 1 bread roll, 1 cheese slice and 1 ham slice. He has bought bread rolls in packs of 9, cheese slices in packs of 6 and ham slices in packs of 8.

When Evan has finished making the sandwiches, there is nothing left in any of the packs. What is the minimum number of packs of each ingredient that Evan could have bought?

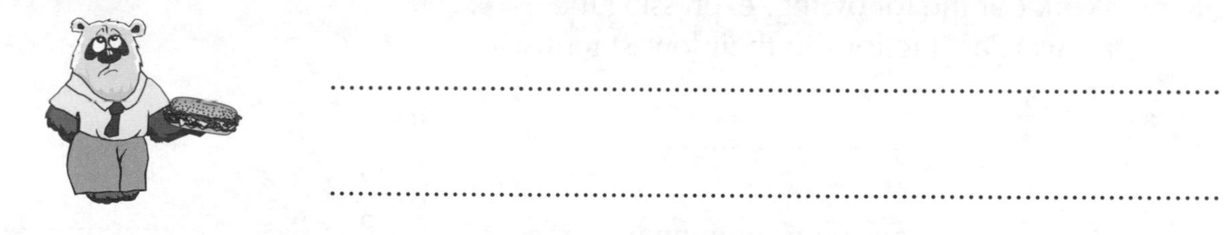

Q8 Sophie has bought some plants — 36 violas, 54 sweet peas and 72 fuchsias. She wants to plant them in pots so that there is only one type of plant in each pot, and each pot has the same number of plants in it.

a) What is the maximum number of plants she could have in each pot?

....

b) How many pots would she need if she put the maximum number of plants in each pot?

....

Fractions Without a Calculator

Fractions could come up on your non-calculator paper, so give these next two pages a go <u>without using your calculator</u>.

Q1 Shade in the correct number of sections to make these diagrams equivalent.

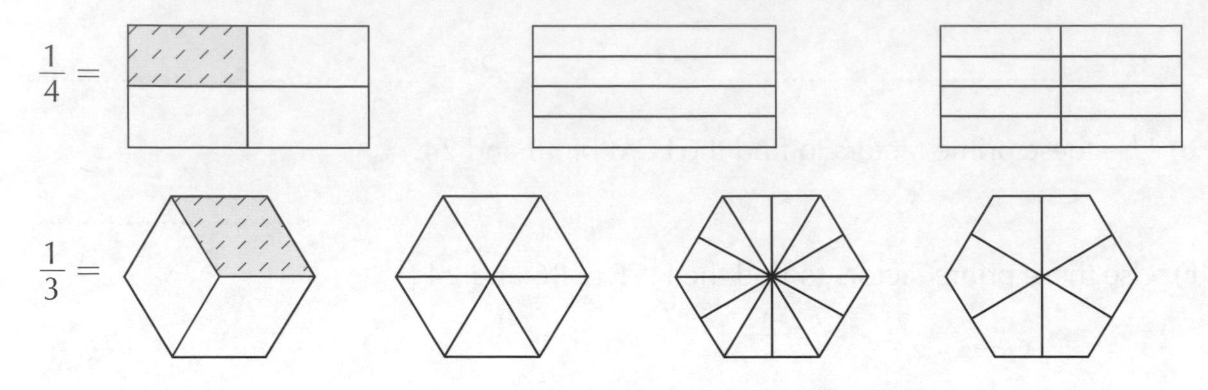

Q2 Write in the missing numbers to cancel these fractions to their simplest forms.

a) $\frac{4}{16} = \frac{1}{\square}$ **b)** $\frac{9}{12} = \frac{3}{\square}$ **c)** $\frac{2}{6} = \frac{\square}{3}$

d) $\frac{8}{12} = \frac{2}{\square}$ **e)** $\frac{6}{18} = \frac{1}{\square}$ **f)** $\frac{24}{32} = \frac{3}{\square}$

Q3 Change these improper fractions to mixed numbers:

a) $\frac{3}{2} =$ .... **b)** $\frac{7}{4} =$ .... **c)** $\frac{8}{3} =$ ....

Q4 Change these mixed numbers to improper fractions:

a) $2\frac{1}{2} =$ .... **b)** $3\frac{1}{3} =$ .... **c)** $1\frac{3}{5} =$ ....

Q5 Work out the following, expressing the answers as fractions in their lowest terms:

a) $\frac{4}{3} \times \frac{3}{4}$ .... **b)** $\frac{1}{4} \div \frac{3}{8}$ ....

c) $\frac{1}{9} \div \frac{2}{3}$ .... **d)** $\frac{2}{5} \times \frac{3}{4}$ ....

e) $\frac{11}{9} \times \frac{6}{5}$ .... **f)** $\frac{15}{24} \div \frac{6}{5}$ ....

When you're multiplying or dividing mixed numbers, you'll need to change them into improper fractions — that's when the number on the top is bigger.

Q6 Work out the following. Express your answer as a mixed number where appropriate.

a) $2\frac{1}{2} \times \frac{3}{5}$ .... **b)** $1\frac{1}{2} \div \frac{5}{12}$ ....

c) $10\frac{4}{5} \div \frac{9}{10}$ .... **d)** $3\frac{7}{11} \div 1\frac{4}{11}$ ....

e) $10\frac{2}{7} \times \frac{7}{9}$ .... **f)** $2\frac{1}{6} \times 3\frac{1}{3}$ ....

Fractions Without a Calculator

Q7 Write in the missing numbers to make the fractions in each list equivalent.

a) $\frac{1}{2} = \frac{2}{\square} = \frac{\square}{6} = \frac{\square}{8} = \frac{5}{10} = \frac{25}{\square} = \frac{\square}{70} = \frac{\square}{100}$

b) $\frac{200}{300} = \frac{100}{\square} = \frac{\square}{15} = \frac{40}{\square} = \frac{120}{180} = \frac{\square}{9} = \frac{\square}{3}$

c) $\frac{7}{10} = \frac{14}{\square} = \frac{\square}{30} = \frac{210}{\square} = \frac{49}{\square} = \frac{\square}{40}$

To make an equivalent fraction, you've got to multiply or divide the top (numerator) and bottom (denominator) by the same thing.

no calculators!!

Q8 Write these sets of fractions in order of size, starting with the smallest:

a) $\frac{1}{5}, \frac{3}{10}$ ....

b) $\frac{3}{7}, \frac{6}{21}$ ....

c) $\frac{11}{15}, \frac{4}{6}, \frac{4}{5}$ ....

d) $\frac{1}{3}, \frac{5}{12}, \frac{4}{6}$ ....

Q9 Evaluate the following, giving your answers as fractions in their lowest terms:

a) $\frac{7}{8} + \frac{3}{8}$ ....

b) $\frac{1}{12} + \frac{3}{4}$ ....

c) $\frac{1}{3} + \frac{3}{4}$ ....

d) $\frac{11}{4} - \frac{2}{3}$ ....

e) $10 - \frac{2}{5}$ ....

f) $8 - \frac{1}{8}$ ....

Q10 Evaluate the following, giving your answers as mixed numbers:

a) $1\frac{2}{5} + 2\frac{2}{3}$ ....

b) $4\frac{2}{3} - \frac{7}{9}$ ....

c) $1\frac{3}{10} + \frac{2}{5}$ ....

d) $3\frac{1}{2} - \frac{2}{3}$ ....

e) $\frac{1}{6} + 4\frac{1}{3}$ ....

f) $1\frac{3}{4} - \frac{1}{5}$ ....

Q11 Calculate the following fractions without using a calculator:

e.g. $\frac{2}{5}$ of 50 = $\frac{2}{5} \times 50 = (50 \div 5) \times 2 = 20$

a) $\frac{1}{8}$ of 32 .... ÷ 8 = ...

b) $\frac{1}{12}$ of 144 .... ÷ .... = ...

c) $\frac{4}{5}$ of 25 (25 ÷ ....) × 4 = ...

d) $\frac{7}{9}$ of 63 (.......... ÷ ....) × 7 = ...

e) $\frac{3}{10}$ of 100 (.......... ÷ ....) × .... = ...

f) $\frac{3}{11}$ of 264 (.......... ÷ ....) × .... = ...

To find a fraction of a number — divide the number by the denominator and then multiply by the numerator.

Fraction Problems

These questions are similar to the ones on the last two pages — just a bit more wordy. Remember to underline the important stuff and work out what the question is actually asking.

Try these questions **without a calculator**, giving any fractions in their simplest form.

Q1 A return car journey from Lancaster to Stoke uses $\frac{5}{6}$ of a tank of petrol. How much does this cost, if it costs £54 for a full tank of petrol?

....

Q2 What fraction of 1 hour is:

a) 5 minutes? .... **b)** 15 minutes? .... **c)** 40 minutes? ....

Q3 If a TV programme lasts 40 minutes, what fraction of the programme is left after:

a) 10 minutes? .... **b)** 15 minutes? .... **c)** 35 minutes? ....

Q4

A café employs eighteen girls and twelve boys to wait at tables. Another six boys and nine girls work in the kitchen.

a) What fraction of the kitchen staff are girls? ....

b) What fraction of the employees are boys? ....

Q5 In the diagram on the right, the area of the circle is $\frac{2}{3}$ the area of the square. The area of the triangle is $\frac{1}{5}$ the area of the circle.

a) Work out $\frac{2}{3} \times \frac{1}{5}$ ....

b) What fraction of the square is shaded? ....

Use a **calculator** for the questions below.

Q6 If I pay my gas bill within seven days, I get a reduction of an eighth of the price. If my bill is £120, how much can I save? ....

Q7 The amount of sugar in Healthybix cereal has been reduced by $\frac{1}{4}$. There is now 4.8 g of sugar in every 50 g of cereal. Work out the original amount of sugar in 50 g of the cereal.

.... g

Q8 Owen gives money to a charity directly from his wages, before tax. One sixth of his monthly earnings goes to the charity, then one fifth of what's left gets deducted as tax. How much money goes to charity and on tax, and how much is left, if he earns £2400?

Charity ...., Tax ...., Left over ....

Fractions, Decimals and Percentages

Fractions, decimals and percentages are different ways of saying "a bit of" something. Make sure you can convert between all of them, they're likely to be in your exam.

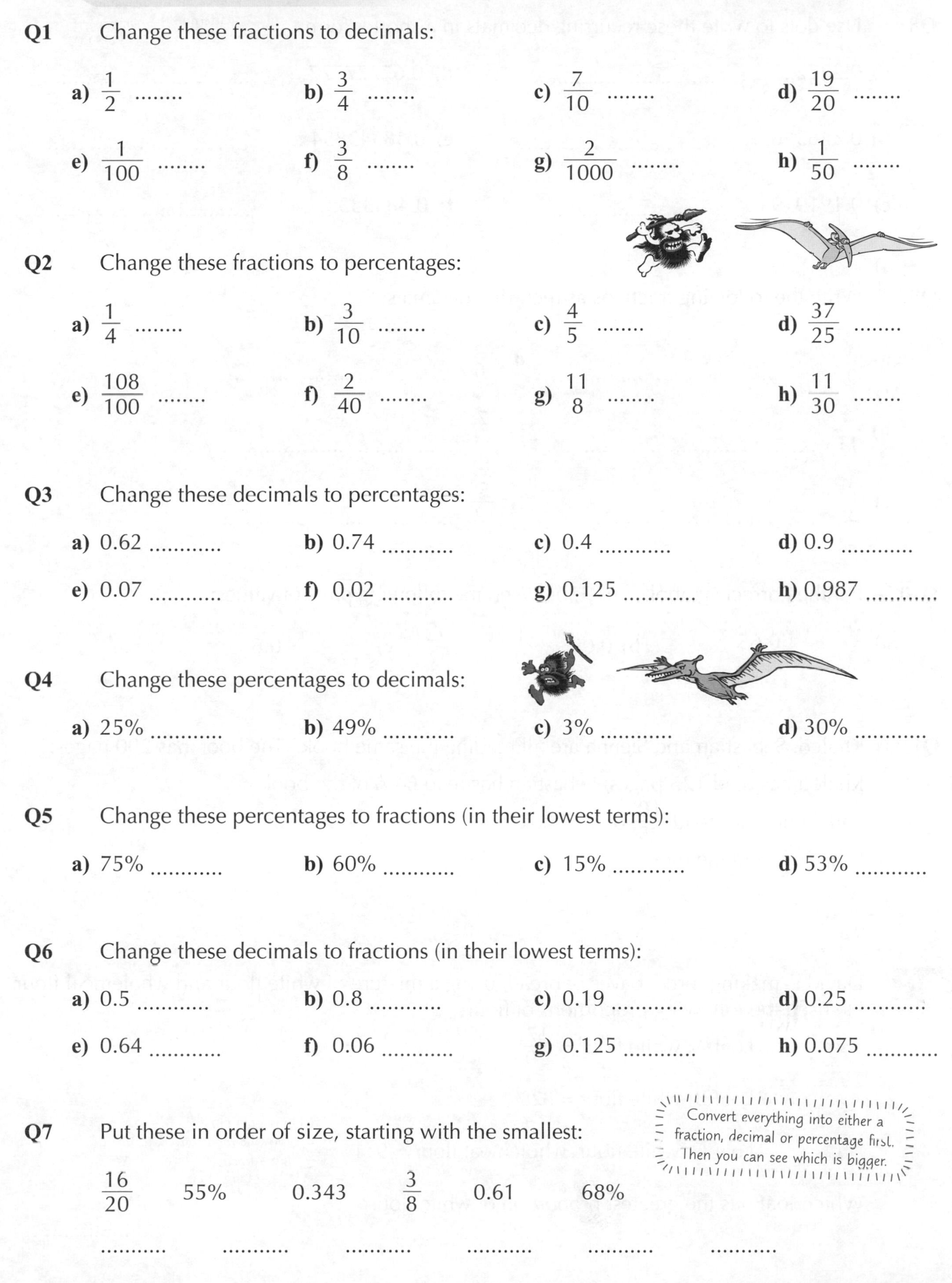

Q1 Change these fractions to decimals:

a) $\frac{1}{2}$ .... **b)** $\frac{3}{4}$ .... **c)** $\frac{7}{10}$ .... **d)** $\frac{19}{20}$ ....

e) $\frac{1}{100}$ .... **f)** $\frac{3}{8}$ .... **g)** $\frac{2}{1000}$ .... **h)** $\frac{1}{50}$ ....

Q2 Change these fractions to percentages:

a) $\frac{1}{4}$ .... **b)** $\frac{3}{10}$ .... **c)** $\frac{4}{5}$ .... **d)** $\frac{37}{25}$ ....

e) $\frac{108}{100}$ .... **f)** $\frac{2}{40}$ .... **g)** $\frac{11}{8}$ .... **h)** $\frac{11}{30}$ ....

Q3 Change these decimals to percentages:

a) 0.62 .... **b)** 0.74 .... **c)** 0.4 .... **d)** 0.9 ....

e) 0.07 .... **f)** 0.02 .... **g)** 0.125 .... **h)** 0.987 ....

Q4 Change these percentages to decimals:

a) 25% .... **b)** 49% .... **c)** 3% .... **d)** 30% ....

Q5 Change these percentages to fractions (in their lowest terms):

a) 75% .... **b)** 60% .... **c)** 15% .... **d)** 53% ....

Q6 Change these decimals to fractions (in their lowest terms):

a) 0.5 .... **b)** 0.8 .... **c)** 0.19 .... **d)** 0.25 ....

e) 0.64 .... **f)** 0.06 .... **g)** 0.125 .... **h)** 0.075 ....

Q7 Put these in order of size, starting with the smallest:

Convert everything into either a fraction, decimal or percentage first. Then you can see which is bigger.

$\frac{16}{20}$ 55% 0.343 $\frac{3}{8}$ 0.61 68%

....

Fractions, Decimals and Percentages

Ugh, decimals that go on forever — that doesn't sound fun...

Look for the repeating pattern in the numbers...

Q8 Use dots to write these recurring decimals in a shorter form:

a) 0.2222... ....

b) 0.346346... ....

c) 0.191919... ....

d) 0.67777777... ....

e) 0.38543854... ....

f) 0.4833333... ....

Q9 Write the following fractions as recurring decimals:

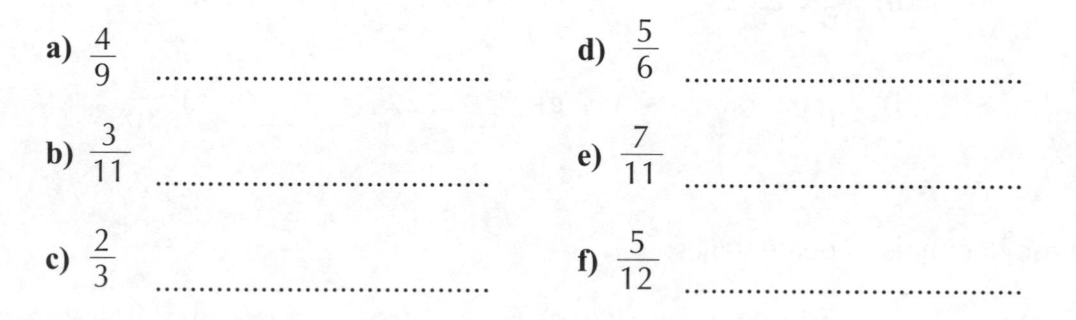

a) $\frac{4}{9}$ ....

b) $\frac{3}{11}$ ....

c) $\frac{2}{3}$ ....

d) $\frac{5}{6}$ ....

e) $\frac{7}{11}$ ....

f) $\frac{5}{12}$ ....

Q10 Put the correct symbol, < or >, between the following pairs of values:

a) $\frac{2}{3}$ 0.65 **b)** 0.09 $\frac{1}{9}$ **c)** $\frac{7}{12}$ 0.6

Q11 Khaled, Sebastian and Sienna are all reading the same book. The book has 200 pages. Khaled has read 125 pages, Sebastian has read 64% of the book and Sienna has read $\frac{18}{25}$ of the book.
Who has read the most?

....

Q12 David is making three loaves of bread, using a mixture of white flour and wholemeal flour. He used the following proportions of flour:

Loaf A: white flour = $\frac{17}{20}$

Loaf B: white flour = 82%

Loaf C: white flour : wholemeal flour = 9 : 1

Turn the ratio into a fraction.

Which loaf has the greatest proportion of white flour?

....

Rounding Numbers

Q1 Round the following to the nearest whole number:

a) 2.9 .... **b)** 26.8 .... **c)** 2.24 ....

d) 11.11 .... **e)** 6.347 .... **f)** 43.5 ....

g) 9.99 .... **h)** 0.41 ....

Nearest <u>whole number</u> means you look at the digit after the decimal point to decide whether to round up or down.

Q2

An average family has 1.7 children. How many children is this to the nearest whole number?

....

Q3 Give these amounts to the nearest pound:

a) £4.29 .... **b)** £16.78 .... **c)** £12.06 ....

d) £7.52 .... **e)** £0.93 .... **f)** £14.50 ....

g) £7.49 .... **h)** £0.28 .... **i)** £9.96 ....

Q4 Round these numbers to the nearest 10:

a) 23 .... **b)** 78 .... **c)** 65 .... **d)** 99 ....

e) 118 .... **f)** 243 .... **g)** 958 .... **h)** 1056 ....

Q5 Round these numbers to the nearest 100:

a) 627 .... **b)** 791 .... **c)** 199 .... **d)** 450 ....

e) 1288 .... **f)** 3329 .... **g)** 2993 ....

Q6 Crowd sizes at sports events are often given exactly in newspapers. Round these exact crowd sizes to the nearest 1000.

a) 23 324 ....

b) 36 844 ....

c) 49 752 ....

Rounding Numbers

Q7 Round the following to 2 d.p.:

a) 17.363 ....
b) 38.057 ....
c) 0.735 ....
d) 5.99823 ....
e) 4.297 ....
f) 7.0409 ....

Q8 Now round these to 3 d.p.:

a) 6.3534 ....
b) 81.64471 ....
c) 0.0075 ....
d) 53.26981 ....
e) 754.39962 ....
f) 0.000486 ....

Q9 Seven people have a meal in a restaurant. The total bill comes to £60. If they share the bill equally, how much should each of them pay?
Round your answer to 2 d.p.

....

Q10 Round these numbers to 1 significant figure.

a) 12 ....
b) 530 ....
c) 1379 ....
d) 0.021 ....
e) 1829.62 ....
f) 0.296 ....

Q11 Round these numbers to 3 significant figures.

a) 1379 ....
b) 1329.62 ....
c) 0.2963 ....
d) 0.02139 ....

The 1st significant figure of any number is the first digit which isn't zero.

Q12 Calculate $\frac{3.2 \times 6.4}{8 \times 0.5}$, giving your answer to 1 d.p.

....

Q13 Calculate $\frac{17.8 - 9.32}{5}$, giving your answer to 3 s.f.

....

Q14 Give these amounts to the nearest hour:

a) 2 hours 12 minutes ....
b) 36 minutes ....
c) 12 hours 12 minutes ....
d) 29 minutes ....
e) 100 minutes ....
f) 90 minutes ....

Estimating

Round to nice easy convenient numbers, then use them to do the calculation. Easy peasy.

Q1 Estimate the answers to these questions.

For example: $12 \times 21 \approx 10 \times 20 = 200$

a) $18 \times 12 \approx$ .... × .... = ...

b) $57 \times 46 \approx$ .... × .... = ...

c) $98 \times 105 \approx$ .... × .... = ...

d) $22 \div 5.3 \approx$ .... ÷ .... = ...

e) $901 \div 33 \approx$ .... ÷ .... = ...

f) $1007 \div 498 \approx$ .... ÷ .... = ...

Q2 Andy earns a salary of £24 108 each year.

a) Estimate how much Andy earns per month. £ ....

b) Andy is hoping to get a bonus this year. The bonus is calculated as £1017 plus 9.7% of each worker's annual salary. Estimate the amount of Andy's bonus.

£ ....

c) Andy pays 10.14% of his regular salary into a pension scheme. Estimate how much of his salary he has left per year after making his pension payments.

£ ....

Q3 Estimate the answers to these calculations:

a) $\dfrac{58.3 \times 11.4}{5.1} \approx$ ....

b) $\dfrac{33.7 \times 6.2}{2.3} \approx$ ....

c) $\dfrac{2.9 \times 48.3}{0.476} \approx$ ....

d) $\dfrac{5.4 \times 8.32}{0.139} \approx$ ....

Q4 Estimate the value of:

a) $\dfrac{\sqrt{3.58 \times 9.14}}{9.37 - 5.84} \approx$ ....

b) $\sqrt{\dfrac{76.31 + 1.41}{2.83 \times 3.25}} \approx$ ....

Q5 Keimia grows and sells kiwis. She sells 206 kiwis in January, 252 kiwis in February and 161 kiwis in March. The kiwis are priced by weight, with an average price of £0.22 per kiwi. Keimia uses the following calculation to estimate how much money she has made:

(Total number of kiwis) × (Price per kiwi) = (200 + 250 + 150) × (0.20) = £120

Is Keimia's estimate likely to be higher or lower than the actual amount of money she made? Explain your answer.

....

Rounding Errors

Q1 Estimate the height of the sunflower next to the adult in the picture.

....

If you have trouble estimating the height by eye, try measuring the man against your finger. Then see how many times that bit of finger fits into the height of the sunflower.

When a number has been rounded off to the nearest unit, the <u>actual number</u> could be up to <u>half a unit</u> bigger or smaller...

Q2 The number of drawing pins in the box has been rounded to the nearest 10.

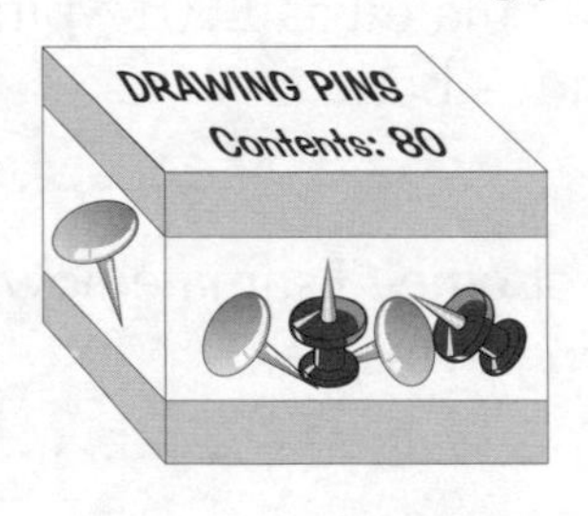

What is the least possible number of drawing pins in the box?

....

Q3 A rectangular rug is 2 metres long and 1 metre wide. Both measurements are given correct to the nearest 10 centimetres.

a) Find the interval in which the actual length of the rug, l, lies. Give your answer in cm.

....

b) State the maximum possible width of the rug in cm.

....

Q4 Matt is buying skirting board to fit to a wall which measures 540 cm to 2 s.f. What length of board should Matt buy to guarantee he has enough for the whole wall?

....

Q5 **a)** The height, h, of the Eiffel Tower is 300.6 m truncated to 1 decimal place. Express the possible values of the height as an inequality.

Remember — truncated means 'cut-off', NOT rounded.

....

b) Neil measures the width of the base of the Eiffel Tower. Neil's measurement is 125.25 m, truncated to 2 d.p. What is the minimum possible value of Neil's actual measurement?

....

Powers

Powers are a way of writing numbers in shorthand — which is quite handy with big numbers. Imagine writing out 2^{138} — 2 × 2 ×... × 2 ×... × 2 × yawn × zzz...

Q1 Complete the following:

a) $2^4 = 2 \times 2 \times 2 \times 2 =$ ....

b) $3^5 = 3 \times$ .... = ...

c) $10^6 = 10 \times$ .... = ...

Q2 Simplify the following using powers:

a) $2 \times 2 \times 2 \times 2 \times 2 \times 2 \times 2 \times 2 =$ ....

b) $12 \times 12 \times 12 \times 12 \times 12 =$ ....

c) $m \times m \times m =$ ....

d) $y \times y \times y \times y =$ ....

Q3 Use your calculator to find the exact value of:

a) 12^5 ....

b) 13^3 ....

c) 8.2^3 ....

d) 0.5^5 ....

Q4 Use the power rules to fill in the blanks (the first two have been done for you):

a) $5^3 \times 5^2 = 5^{3+2} = 5^5$

b) $9^{17} \div 9^4 = 9^{17-4} = 9^{13}$

c) $6^3 \times 6^5 =$ .... = ...

d) $12^8 \div 12^2 =$ .... = ...

e) $7^{22} \div 7^{15} =$ .... = ...

f) $13^6 \times 13 =$ .... = ...

Remember, $13 = 13^1$.

g) $3^{14} \div 3^9 =$ .... = ...

h) $8^{12} \times 8^3 =$ .... = ...

Q5 Use the power rules to complete the following (the first one has been done for you):

a) $(15^3)^2 = 15^{3\times2} = 15^6$

b) $(3^8)^4 =$ .... = ...

c) $(21^2)^6 =$ .... = ...

d) $(17^5)^8 =$ .... = ...

Q6 Without using a calculator, work out the following:

a) $\dfrac{5^8 \times 5^3}{5^3 \times 5^6}$ ....

b) $\dfrac{7^0}{7^1 \times 7^1}$ ....

Q7 Complete the following:

For negative powers — just turn the whole thing the other way up and change the power to a positive. Magic.

For example $3^{-3} = \dfrac{1}{3^3} = \dfrac{1}{27}$

a) $2^{-5} =$ .... = ...

b) $8^{-2} =$ .... = ...

c) $\left(\dfrac{3}{2}\right)^{-2} =$ .... = ... = ...

d) $\left(\dfrac{3}{5}\right)^{-3} =$ .... = ... = ...

Roots

Don't let the fancy symbols put you off — roots aren't so bad really. For example, a square root just means "What number times by itself gives...?". And remember — numbers always have two square-roots — a positive one and a negative one.

Q1 Use the √ button on your calculator to find the following positive square roots to the nearest whole number.

a) $\sqrt{60} \approx$ ....
b) $\sqrt{19} \approx$ ....
c) $\sqrt{34} \approx$ ....
d) $\sqrt{200} \approx$ ....
e) $\sqrt{520} \approx$ ....
f) $\sqrt{75} \approx$ ....
g) $\sqrt{750} \approx$ ....
h) $\sqrt{0.9} \approx$ ....
i) $\sqrt{170} \approx$ ....
j) $\sqrt{7220} \approx$ ....
k) $\sqrt{1\,000\,050} \approx$ ....
l) $\sqrt{27} \approx$ ....

Q2 Without using a calculator, write down both square roots of each of the following:

a) 4 ....
b) 16 ....
c) 9 ....
d) 49 ....
e) 25 ....
f) 100 ....
g) 144 ....
h) 64 ....
i) 81 ....

Q3 Use your calculator to find the following:

a) $\sqrt[3]{4096} =$ ....
b) $\sqrt[3]{1728} =$ ....
c) $\sqrt[3]{1331} =$ ....
d) $\sqrt[3]{1\,000\,000} =$ ....
e) $\sqrt[3]{1} =$ ....
f) $\sqrt[3]{0.125} =$ ....

Q4 Without using a calculator, find the value of the following:

a) $\sqrt[3]{64} =$ ....
b) $\sqrt[3]{27} =$ ....
c) $\sqrt[3]{1000} =$ ....
d) $\sqrt[3]{8} =$ ....

Q5 Use your calculator to find the following roots:

a) $\sqrt[4]{6561} =$ ....
b) $\sqrt[7]{2187} =$ ....
c) $\sqrt[5]{7776} =$ ....
d) $\sqrt[9]{512} =$ ....

Don't forget — to find the volume of a cube, you multiply the length by itself and by itself again.

Q6 Nida is buying a small storage box online. She sees a cube box with volume of 125 cm^3. What is the length of each box edge?

....

Q7 A farmer is buying fencing to surround a square field of area 3600 m^2. What length of fencing does he need to buy?

....

Q8 Use your calculator to find the value of the following. Give your answers to 2 d.p.

a) $\sqrt[3]{458} + 16 \times 2.8 =$ ....
b) $7.5^2 - \sqrt{192} \div 2 =$ ....

Standard Form

Writing very big (or very small) numbers gets a bit messy with all those zeros if you don't use this standard form. But of course, the main reason for knowing about standard form is... you guessed it — it's in the exam.

Q1 Write these numbers in standard form:

a) 36 700 = ...

b) 0.054 = ...

c) 194 000 = ...

d) 280 = ...

e) 0.000811 = ...

f) 0.0000792 = ...

Q2 Write the following as ordinary numbers:

a) 3.86×10^4 = ...

b) 5.1×10^{-5} = ...

c) 2.62×10^6 = ...

d) 1.37×10^{-3} = ...

e) 6.05×10^3 = ...

f) 4.621×10^{-4} = ...

Q3 This table gives the diameter and distance from the Sun of some planets.

From the table write down which planet is:

a) smallest in diameter ....

b) largest in diameter ....

c) nearest to the Sun ....

d) furthest from the Sun ....

Planet	Distance from Sun (km)	Diameter (km)
Earth	1.5×10^8	1.3×10^4
Mercury	5.81×10^7	4.9×10^3
Jupiter	7.8×10^8	1.4×10^5
Neptune	4.52×10^9	4.9×10^4

To × or ÷ in standard form, rearrange to put the front numbers and powers of 10 together, then multiply/divide both. For + or –, add/subtract the front numbers only (but check the powers of 10 are the same).

Q4 Evaluate the following without using a calculator.
Write your answers in standard form.

a) $(3 \times 10^4) \times (2 \times 10^2)$ ....

b) $(2.8 \times 10^5) \div (1.4 \times 10^3)$ ....

c) $(5 \times 10^5) + (3.6 \times 10^5)$ ....

d) $(5.2 \times 10^7) - (1.8 \times 10^7)$ ....

Q5 Use a calculator to work out the following. Write your answers in standard form.

a) $(3.6 \times 10^8) \div (4.5 \times 10^5)$ = ...

b) $(2.5 \times 10^3) \times (9.8 \times 10^{-9})$ = ...

c) $(7.68 \times 10^8) \div (6.4 \times 10^{12})$ = ...

d) $(3.2 \times 10^{-5}) \times (5.5 \times 10^{11})$ = ...

Simplifying

Algebra can be pretty scary at first. But don't panic — to simplify an expression, just remember to collect like terms. Eventually you'll be able to do it without thinking, just like riding a bike. But a lot more fun, obviously...

Q1 By collecting like terms, simplify the following. The first one is done for you.

a) $6x + 3x - 5 = 9x - 5$

b) $2x + 3x - 5x =$ ....

c) $9f + 16f + 15 - 30 =$ ....

d) $14x + 12x - 1 + 2x =$ ....

e) $3x + 4y + 12x - 5y =$ ....

f) $11a + 6b + 24a + 18b =$ ....

g) $9f + 16g - 15f - 30g =$ ....

h) $14a + 12a^2 - 3 + 2a =$ ....

Q2 Simplify the following. The first one is done for you.

a) $3x^2 + 5x - 2 + x^2 - 3x = 4x^2 + 2x - 2$

b) $5x^2 + 3 + 3x - 4 =$ ....

c) $13 + 2x^2 - 4x + x^2 + 5 =$ ....

d) $7y - 4 + 6y^2 + 2y - 1 =$ ....

e) $2a + 4a^2 - 6a - 3a^2 + 4 =$ ....

f) $15 - 3x - 2x^2 - 8 - 2x - x^2 =$ ....

g) $x^2 + 2x + x^2 + 3x + x^2 + 4x =$ ....

h) $2y^2 + 10y - 7 + 3y^2 - 12y + 1 =$ ....

Multiply/divide any numbers first, then deal with the letters. You'll need to use your power rules if the letters are the same.

Q3 Simplify the following expressions.

a) $b \times b \times b \times b \times b =$ ....

b) $c \times d \times 4 =$ ....

c) $2e \times 6f =$ ....

d) $5g \times 3g^2 =$ ....

e) $7h^2 \times 8 =$ ....

f) $2j^2 \times j \times k =$ ....

g) $4p^2 \div 6p =$ ....

h) $10n^3 \div 5n^2 =$ ....

Q4 Simplify the following.

a) $5 + 2 + 2\sqrt{3} + \sqrt{3} =$ ....

b) $2 + \sqrt{2} + 1 + \sqrt{2} =$ ....

c) $6 + 2\sqrt{5} - 2 + 3\sqrt{5} =$ ....

d) $10 - 3\sqrt{6} - 7 + 2\sqrt{6} =$ ....

Multiplying Out Brackets

When you're multiplying out brackets, you have to multiply everything outside the bracket by everything inside it.

Q1 Expand the following brackets.
The first one is done for you.

a) $2(x + y) = 2x + 2y$

b) $4(x - 3)$ = ...

c) $8(x^2 + 2)$ = ...

d) $-2(x + y)$ = ...

e) $-3(y + 4)$ = ...

f) $5(2 - y)$ = ...

g) $x(x + 8)$ = ...

h) $3x(x + y)$ = ...

Careful with the minus signs — they multiply both terms in the bracket.

Q2 Multiply out the brackets and simplify where possible:

a) $8(a + b) + 2(a + 2b)$ = ...

b) $4(x - 2y) + 5(x + 3y)$ = ...

c) $a - 4(a + b)$ = ...

d) $e(e + 2f) + 3e(e - f)$ = ...

e) $4x(x + 2) - 2x(3 - x)$ = ...

f) $3(2 + ab) + 5(1 - ab)$ = ...

g) $2(x - 2y) - 2x(x + z)$ = ...

h) $x^2(x + 1)$ = ...

When you've got two brackets multiplied together, always use FOIL — First, Outside, Inside, Last.

Q3 Multiply out the brackets and simplify your answers where possible:

a) $(x + 1)(x + 2)$ = ...

b) $(x - 3)(x + 5)$ = ...

c) $(x + 10)(x + 3)$ = ...

d) $(x - 5)(x - 2)$ = ...

e) $(4 - x)(7 - x)$ = ...

f) $(2x + 1)(x + 3)$ = ...

g) $(3x + 2)(2x - 4)$ = ...

h) $(2 + 3x)(3x - 1)$ = ...

Remember to write squared brackets as two separate brackets.

Q4 Multiply out these squared brackets:

a) $(x - 1)^2$ = ...

b) $(x + 3)^2$ = ...

c) $(2x - 5)^2$ = ...

d) $(4x + 3)^2$ = ...

Factorising

FACTORISING — it's just putting the brackets back in. All you have to do is look for common factors that will go into every term. The factors can be numbers or letters.

Q1 Factorise the expressions below. Each has 4 as a common factor.

a) $4x + 8$ = ...
b) $12 - 8x$ = ...
c) $4 - 16x$ = ...
d) $24x + 28$ = ...
e) $32 - 20x$ = ...
f) $4x^2 + 64$ = ...

Q2 Factorise the expressions below. Each has x as a common factor.

a) $2x + x^2$ = ...
b) $2x - x^2$ = ...
c) $x^2 - 7x$ = ...
d) $x - 16x^2$ = ...
e) $4x^2 - 3x$ = ...
f) $6x + 13x^2$ = ...

Q3 Factorise the expressions below by taking out any common factors.

a) $2x + 4$ = ...
b) $3x + 12$ = ...
c) $24 + 12x$ = ...
d) $16x + 4y$ = ...
e) $3x + 15$ = ...
f) $30 + 10x$ = ...
g) $9x^2 + 3x$ = ...
h) $5x^2 + 10x$ = ...
i) $7x^2 + 21x$ = ...
j) $8x^2 + 4x$ = ...

First look for any numbers the terms have in common, then look for the letters.

Q4 Factorise the expressions below.

a) $3y + xy^2$ = ...
b) $a + 2a^2b$ = ...
c) $4mn^2 + 2m$ = ...
d) $3g^2h - 9g$ = ...

Q5 Using the fact that $a^2 - b^2 = (a + b)(a - b)$, factorise the following expressions:

a) $x^2 - 9$ = ...
b) $y^2 - 16$ = ...
c) $25 - z^2$ = ...
d) $36 - a^2$ = ...

Q6 Factorise the following expressions:

a) $4x^2 - 9$ = ...
b) $9y^2 - 4$ = ...
c) $25 - 16z^2$ = ...
d) $1 - 36a^2$ = ...

Solving Equations

Top Tip

When you're asked to solve an equation, you just have to find the value of the letter in the equation. If the letter is 'x', you need to rearrange the equation to get 'x = a number'.

Q1 Solve these equations:

a) $a + 6 = 20$ ....
b) $b + 12 = 30$ ....
c) $h - 14 = 11$ ....

d) $3 + f = -7$ ....
e) $i - 38 = 46$ ....
f) $k - 6.4 = 2.9$ ....

Q2 Solve these equations:

a) $4m = 28$ ....
b) $15p = 645$ ....
c) $6r = -9$ ....

d) $\frac{t}{5} = 11$ ....
e) $\frac{w}{12} = 9$ ....
f) $\frac{v}{8} = 14$ ....

Q3 Solve these equations:

a) $3x + 2 = 14$ ....
b) $5x - 4 = 31$ ....
c) $8 + 6x = 50$ ....

d) $\frac{x}{3} + 4 = 10$ ....
e) $4 + \frac{x}{9} = 6$ ....
f) $\frac{x}{10} - 11 = 9$ ....

Q4 Solve these equations:

a) $x + 2 = 2x - 3$ ....
b) $2x - 1 = 3x - 4$ ....
c) $5x - 12 = 2x + 6$ ....

d) $10x - 5 = 3x + 9$ ....
e) $7x - 8 = 9x - 16$ ....
f) $11x + 3 = 4x + 10$ ....

Q5 Solve these equations:

a) $2(x + 1) = 6$ ....
b) $5(x - 3) = 2x + 15$ ....
c) $4(x + 5) = 2(3x + 1)$ ....

Q6 Solve these equations:

a) $x^2 = 25$ ....
b) $2y^2 = 32$ ....
c) $8u^2 = 72$ ....

Remember to find both the positive and negative square roots.

Expressions, Formulas and Functions

Don't get tempted to shove the numbers into your calculator all at once for these questions — you're a lot less likely to make a mistake if you do it in stages using BODMAS.

Q1 $m = 5n + 2p$. Find the value of m when $n = 3$ and $p = 4$. ....

Q2 $c = 8d - 4e$. Find the value of c when $d = 6$ and $e = 2$. ....

Q3 The cost of framing a picture, C pence, depends on the dimensions of the picture. If $C = 10L + 5W$, where L is the length in cm and W is the width in cm, then find the cost of framing:

a) a picture of length 40 cm and width 24 cm ....

b) a square picture of sides 30 cm. ....

Q4 To work out the number of sheep to put in each of his fields, a farmer uses the formula:

$$\text{number of sheep in each field} = \frac{\text{total number of sheep} - 8}{\text{number of fields available}}$$

Find the number of sheep the farmer should put in each field if:

a) there are 5 fields available and 243 sheep ....

b) there are 7 fields available and 582 sheep ....

Q5 A baker uses the following formula to work out how long to bake her cakes for:

$$\text{number of minutes} = \frac{\text{surface area of cake (cm}^2\text{)} \times \text{oven temperature (°C)}}{300}$$

Find the number of minutes that the baker should bake a cake for if it has a surface area of 50 cm² and the oven is at a temperature of 180 °C.

....

Q6 Given the formula $s = ut + \frac{1}{2}at^2$, find s when $u = 2$, $t = 3$, and $a = 5$.

....

Remember, – – makes +.

Q7 $u = \sqrt{v^2 - 2as}$. Find the value of u given that $v = 9$, $a = -3.5$ and $s = 4$. Give your answer to 2 d.p.

....

Q8 The function machine below represents the function 'multiply by 3 then subtract 4'.

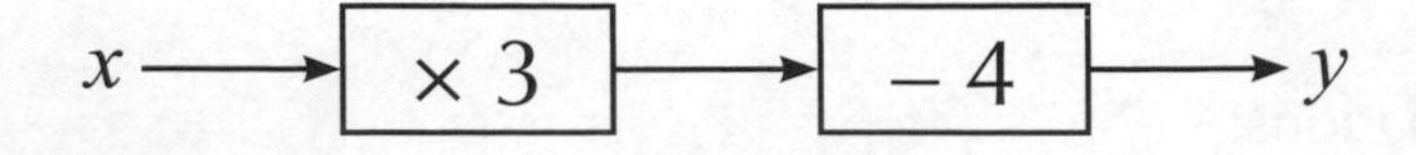

a) Find the value of y when $x = 6$. ....

b) Find the value of x when $y = 29$. ....

Formulas and Equations from Words

It's no big mystery — algebra is just like normal sums, but with the odd letter or two stuck in for good measure.

Q1 There are n books in a pile. Write an expression for the number of books in a pile that has:

a) 3 more books ....

b) 4 fewer books ....

c) Twice as many books ....

Q2 Olivia, Lauren and Ryan take a maths test at school. Olivia gets x marks. Lauren gets twice as many marks as Olivia. Ryan gets 3 more marks than Olivia.

Write a simplified expression for the total number of marks that Olivia, Lauren and Ryan received. ....

Q3 Rebecca has y cats and 4 times as many fish. The sum of the number of cats and the number of fish is 20. How many cats does Rebecca have?

....

Q4 The cost (£C) of hiring a mountain bike is £10, plus £5 for each hour you use the bike (h).

a) Write down a formula that can be used for working out the cost of hiring a bike.

....

b) Tom pays £47.50 to hire a mountain bike. For how long did he use the bike?

....

Q5 The cost (£c) of a taxi journey is £1.40 per mile (m), plus an extra charge of £1.50.

a) Write down a formula with c as the subject. ....

b) Frank wants to know how far his hotel is from the airport. He knows the taxi fare from one to the other is £9.90. How far is the hotel from the airport? ....

Q6 Frances, Millicent and Winston collect thimbles. Frances has x thimbles. Millicent has 3 times as many thimbles as Frances. Winston has 5 fewer thimbles than Millicent.

Between them, they have 23 thimbles. How many thimbles do they each have?

Frances .... Millicent .... Winston ....

Q7 Florence booked 5 tickets for a concert. Each ticket cost £18.50 plus a booking fee. There was also a fee of £5 added to the order to cover postage. The total amount Florence paid was £116. Work out the booking fee for each ticket.

....

Formulas and Equations from Diagrams

I know, I know, this page looks a bit scary, but it really isn't too different from the last one — it's still all about writing and solving your own equations. You've just got the extra fun of shapes thrown in...

Q1 a) Write an expression for the perimeter of this rectangle. ....

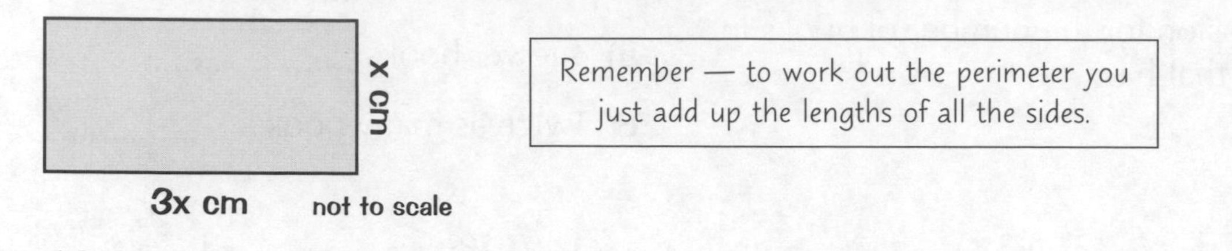

b) If the perimeter of the rectangle is 36 cm, find the value of x. ....

Q2 Find the length of one side of this equilateral triangle. ....

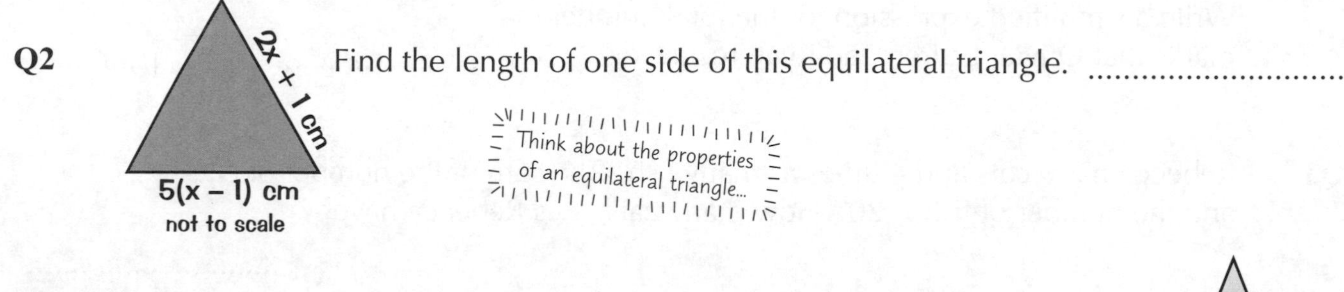

Q3 a) Write a formula for P, the perimeter of this isosceles triangle, in terms of y.

....

b) If the perimeter of the triangle is 13 cm, find the value of y.

....

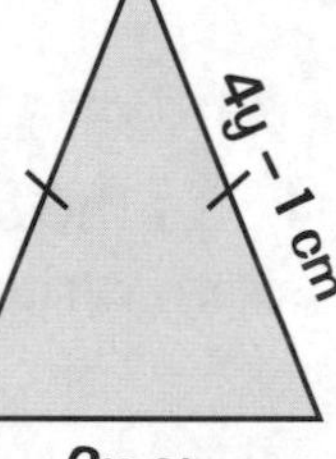

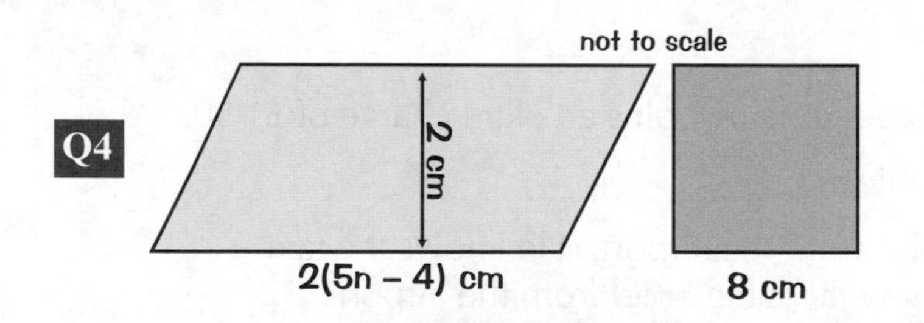

Remember, for a parallelogram:
Area = base × vertical height.

Q4 The area of the square is the same as the area of the parallelogram. Find the value of n.

....

Q5 The perimeter of the rectangle is 2 cm greater than the perimeter of the square. Find the length of one side of the square.

....

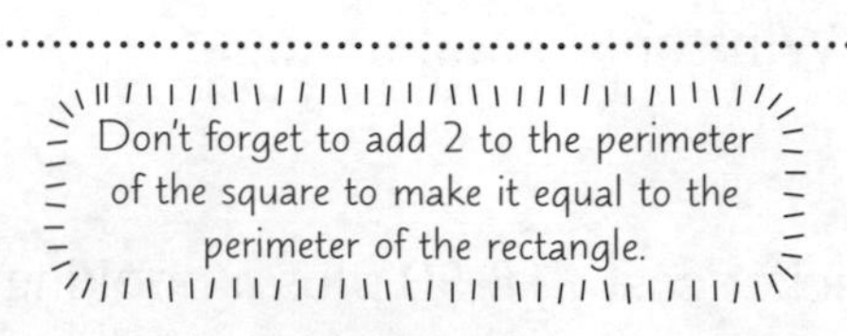

Q6 The areas of these two triangles are equal. Find the value of x.

....

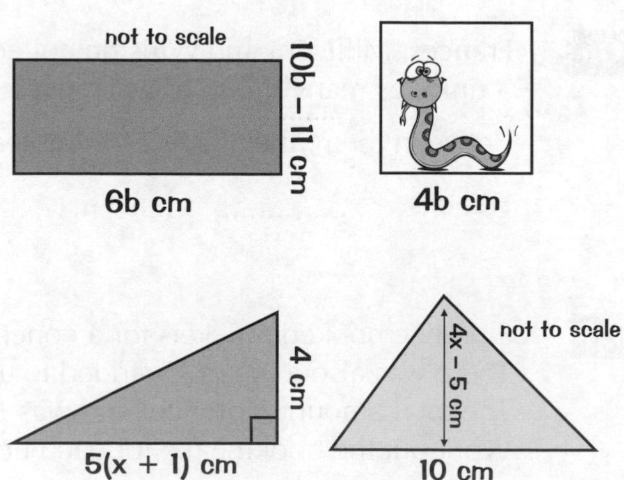

Rearranging Formulas

Rearranging is getting the letter you want out of the formula and making it the subject. And it's the same method as for solving equations, which can't be bad.

Q1 Rearrange the following formulas to make the letter in brackets the new subject:

a) $y = x + 4$ (x) ....

b) $y = 2x + 3$ (x) ....

c) $y = 4x - 5$ (x) ....

d) $a = 7b + 10$ (b) ....

e) $w = 14 + 2z$ (z) ....

f) $s = 4t - 3$ (t) ....

g) $y = 3x + ½$ (x) ....

h) $y = 3 - x$ (x) ....

i) $y = 5(x + 2)$ (x) ....

Q2 Rearrange the following to make the letter in brackets the subject of the formulas:

a) $y = \frac{x}{10}$ (x) ....

b) $s = \frac{t}{14}$ (t) ....

c) $a = \frac{2b}{3}$ (b) ....

d) $d = \frac{3e}{4}$ (e) ....

e) $f = \frac{3g}{8}$ (g) ....

f) $y = \frac{x}{5} + 1$ (x) ....

g) $y = \frac{x}{2} - 3$ (x) ....

h) $a = \frac{b}{3} - 5$ (b) ....

Q3 Rearrange the following to make x the subject of the formula:

a) $y = x^2$ ....

b) $y = 2x^2$ ....

c) $y = x^2 - 3$ ....

d) $y = x^2 + 4$ ....

e) $y = 5x^2$ ....

f) $y = x^2 - 11$ ....

Q4 Rearrange the following to make x the subject of the formula:

a) $y = \frac{1}{x}$ ....

b) $y = \frac{3}{x}$ ....

c) $y = \frac{1}{x} - 5$ ....

d) $y = \frac{1}{x} + 6$ ....

Q5 Rearrange the following to make y the subject of the formula:

a) $x = yz + y$ ....

b) $z = y - ny$ ....

c) $a = 2y - cy$ ....

d) $m = 2py + y$ ....

e) $xy = y + 3$ ....

f) $3ay = 2y + 5$ ....

Sequences

Ahhh, sequences — they're not so bad once you've spotted the pattern. It all comes down to figuring out what you have to do to get from one term to the next...

Q1 Draw the next two pictures in each pattern.

a) How many match sticks are used in each picture?

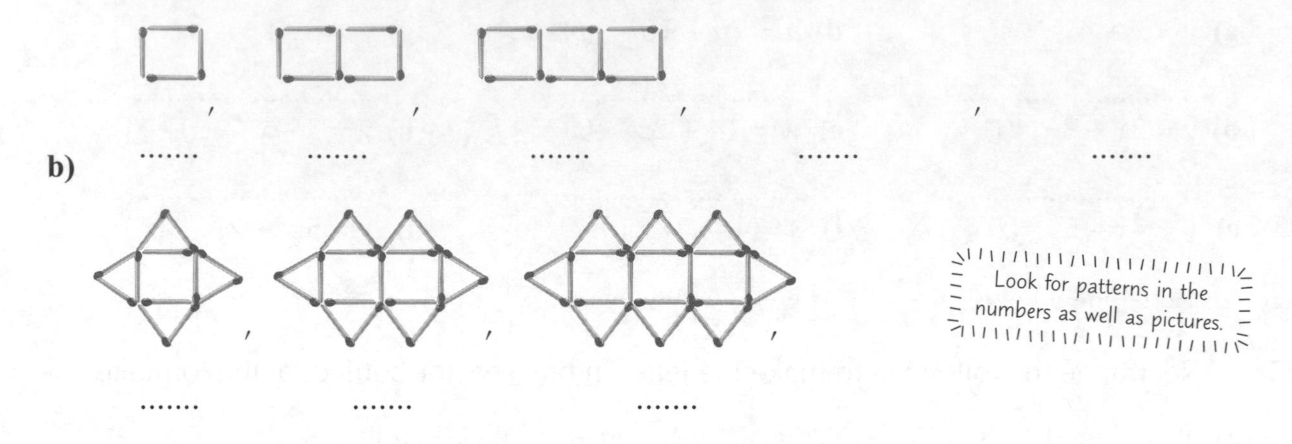

……. , ……. , ……. , ……. , …….

b)

……. , ……. , ……. , ……. , …….

Q2 In each of the questions below, write down the next three numbers in the sequence and write the rule that you used.

a) 1, 3, 5, 7, ………, ………, ……… Rule ………………………………………

b) 16, 8, 4, 2, ………, ………, ……… Rule ………………………………………

c) 3, 30, 300, 3000, ……………, ……………, …………… Rule ………………………

d) 15, 11, 7, 3, ………, ………, ……… Rule ………………………………………

e) 0.1, 0.4, 1.6, 6.4, ……………, ……………, …………… Rule ………………………

Q3 For each of the sequences below, write down the next three terms.

a) 1, 2, 4, 7, 11, ………, ………, ………

b) 2, 5, 10, 17, 26, ………, ………, ………

c) 6, 8, 12, 18, 26, ………, ………, ………

Careful with these ones — you need to spot how the difference between each term is changing...

Q4 a) Draw the next pattern in the sequence below.

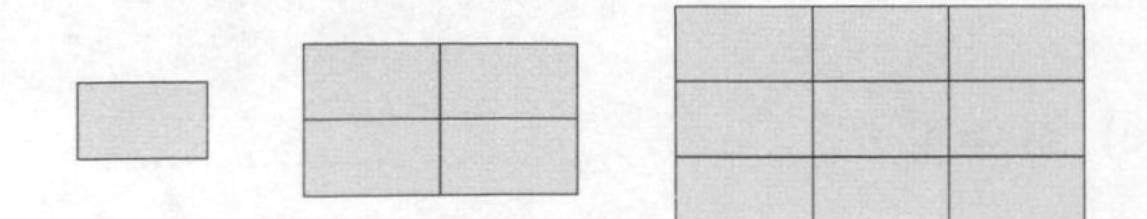

b) How many rectangles would there be in the 6th pattern? ………………………………

Sequences

Q5 In the following, use the rule given to generate (or make) the terms of the sequence.

a) $3n + 1$ so if $n = 1$ the 1st term is $\underline{(3 \times 1) + 1} = \underline{4}$

$n = 2$ the 2nd term is .... = ...

$n = 3$ .... = ...

$n = 4$ .... = ...

b) $5n - 2$, when $n = 1, 2, 3, 4$ and 5 produces the sequence ...., ..., ..., ..., ...

c) n^2, when $n = 1, 2, 3, 4$, and 5 produces the sequence ...., ..., ..., ..., ...

Q6 **3 3 6 9 15 24 ...**

a) What is the rule for this sequence?

....

b) Find the next three terms in the sequence.

....

Q7 The first five terms of a sequence are 3, 7, 11, 15, 19...
Is 34 a term in the sequence? Explain your answer.

....

Q8 A sequence is given by the rule $3n + 7$.

a) Find the 10th term in the sequence.

....

b) Is 53 a term in the sequence?

....

Q9 Write down an expression for the nth term of the following sequences:

a) 2, 4, 6, 8, ...

....

b) 1, 3, 5, 7, ...

....

c) 4, 7, 10, 13, ...

....

The common difference tells you what multiple of n to use in the expression.

Q10 Mike has opened a bank account to save for a holiday. He opened the account with £20 and puts £15 into the account at the end of each week. So at the end of the first week the balance of the account is £35. What will the balance of the account be at the end of:

a) the third week? ....

b) the fifth week? ....

c) Write down an expression that will let Mike work out how much he'll have in his account after any number of weeks he chooses.

....

d) What will the balance of the account be after 18 weeks?

....

Inequalities

Number lines are pretty handy for showing inequalities, but just remember — an <u>open</u> circle is used for $<$ (less than) and $>$ (greater than). And a <u>closed</u> circle is used for $\leq$ (less than or equal to) and $\geq$ (greater than or equal to).

Q1 Write down an inequality for each of the diagrams below.

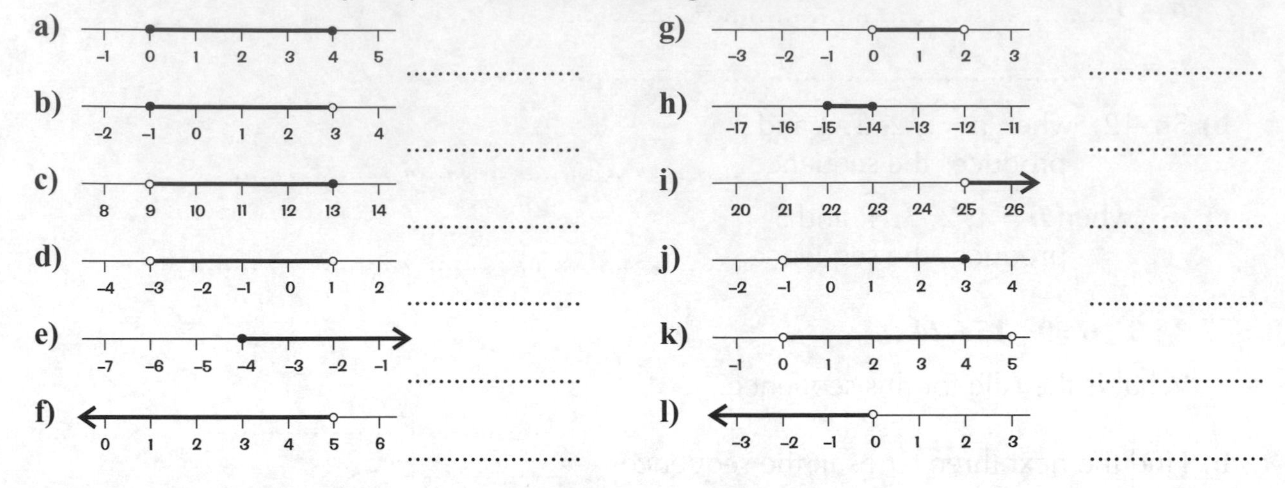

Q2 Show each of the following inequalities on a number line.

a) $x > 5$

b) $-5 < x < 2$

c) $-2 \leq x < 3$

d) $-3 \leq x \leq -2$

Q3 x is an integer such that $-2 \leq x < 4$. Write down all possible values of x.

....

Q4 Solve the following inequalities:

a) $2x \geq 16$ ....

b) $4x > -20$ ....

c) $x + 2 > 5$ ....

d) $x - 3 \leq 10$ ....

e) $10x > -2$ ....

f) $5 + x \geq 12$ ....

g) $\frac{x}{4} > 10$ ....

h) $\frac{x}{3} \leq 1$ ....

i) $16 - 3x > 4$ ....

j) $5x + 7 \leq 32$ ....

k) $3x + 12 \leq 30$ ....

l) $7 - 2x \leq -8$ ....

Q5 Solve the following inequalities:

a) $4x - 7 \geq x + 2$ ....

b) $5x - 2 < 3x + 8$ ....

c) $7x + 10 \leq 5x + 12$ ....

d) $6x - 11 > 4x + 7$ ....

Q6 Find the integer value of x which satisfies both $x - 3 > 2$ and $2x + 1 < 15$.

....

Quadratic Equations

Most of the work involved in solving quadratic equations is the factorising part — the good news is you can check you've done it right by expanding the brackets. If this doesn't give you the expression you started with, just try again...

Q1 Factorise the following:

a) $x^2 + 3x + 2$ ....

b) $x^2 - 7x - 8$ ....

c) $x^2 - 8x + 15$ ....

d) $x^2 - 3x - 18$ ....

e) $x^2 - 5x + 4$ ....

f) $x^2 + 5x + 6$ ....

Q2 Solve these quadratic equations by factorising:

a) $x^2 + 3x - 10 = 0$ ....

b) $x^2 - 5x + 6 = 0$ ....

c) $x^2 - 2x + 1 = 0$ ....

d) $x^2 - 4x + 3 = 0$ ....

e) $x^2 - x - 20 = 0$ ....

f) $x^2 + 14x + 49 = 0$ ....

Q3 Rearrange into the form "$x^2 + bx + c = 0$", then solve by factorising:

a) $x^2 + 6x = 16$ ....

b) $x^2 + 5x = 36$ ....

c) $x^2 + 4x = 45$ ....

d) $x^2 = 5x$ ....

e) $x^2 = 11x$ ....

f) $x^2 - 21 = 4x$ ....

g) $x^2 + 48 = 26x$ ....

h) $x^2 + 36 = 13x$ ....

Careful here — remember lengths have to be positive.

Q4 A rug has length x m and width $(x - 1)$ m.

x – 1 m

x m

a) Write down an expression for the area of the rug.

....

b) If the area of the rug is 6 m², find the value of x.

....

Q5 The area of a rectangular swimming pool is 28 m².
The width is x m. The length is $(x + 3)$ m.
Find the value of x.

....

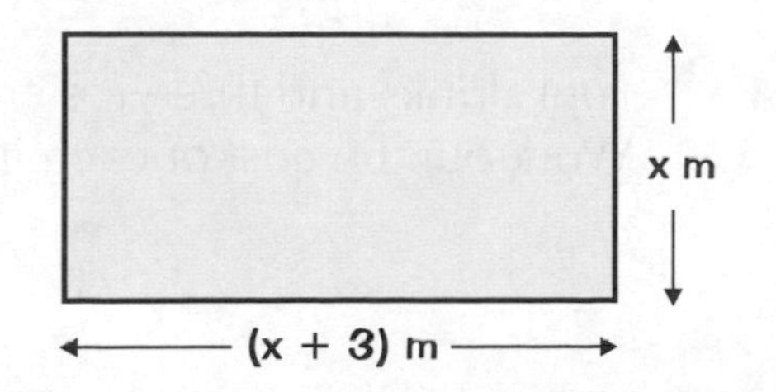

Simultaneous Equations

To solve simultaneous equations from scratch, you've got to get rid of either x or y first — to leave you with an equation with just one unknown in it.

Q1 Solve the following simultaneous equations:

a) $4x + 6y = 16$
$x + 2y = 5$ ....

b) $3x + 8y = 24$
$x + y = 3$ ....

c) $-8x + 3y = 20$
$2x + 3y = 10$ ....

d) $4x + 5y = 12$
$-8x + 4y = 32$ ....

e) $2x + 3y = 9$
$6x - 4y = 14$ ....

f) $8x - 2y = 6$
$5x + 4y = 30$ ....

g) $12x + 3y = 24$
$4x + 5y = -8$ ....

h) $3x - 2y = 18$
$9x - 5y = 60$ ....

Don't forget to substitute your x and y values back into one of the equations at the end to make sure it works. If it doesn't, do it all again until it does...

Q2 Solve each pair of simultaneous equations:

a) $5x + 2y = 16$
$2x + y = 7$ ....

b) $-6x + 7y = 11$
$2x + 2y = 18$ ....

c) $8x + 7y = 20$
$4x - 2y = 32$ ....

d) $6x - 7y = 16$
$3x + 2y = 19$ ....

You'll need to multiply both equations by suitable numbers here to match up the numbers in front of the x's or y's.

Q3 Solve the following simultaneous equations:

a) $3x + 2y = 7$
$2x + 5y = 12$ ....

b) $5x - 2y = 4$
$-4x + 3y = 1$ ....

c) $5x - 4y = 11$
$2x - 9y = -3$ ....

d) $3x + 4y = 10$
$5x - 7y = 3$ ....

e) $-3x + 2y = 1$
$2x + 3y = 21$ ....

f) $4x + 3y = 8$
$6x - 5y = -7$ ....

Q4 Four drinks and three ice creams cost £12. Three drinks and six ice creams cost £16.50. Work out the cost of each item.

Drink: £ .... Ice cream: £ ....

Proof

Proofs look pretty scary but they're usually alright once you get started. For these first few questions you just need to do some rearranging and make one side of the identity look like the other side.

Q1 Prove the following identities:

a) $(n + 5)^2 - (n + 1)^2 \equiv 8(n + 3)$

....

b) $(n - 3)^2 - (n + 1)(n - 1) \equiv 2(5 - 3n)$

....

c) $n(n - 1) + (n - 2)(3 - n) \equiv 2(2n - 3)$

....

You might have to prove that a statement isn't true. The best way to do this is to find an example that doesn't work...

Q2 Maisy says "all prime numbers are odd". Find an example to show that Maisy is wrong.

....

Q3 Timothy says "If the sum of two integers is even, one of the integers must be even". Prove that Timothy is wrong.

....

Q4 For each of the following statements, provide an example to show it is false.

a) The difference between any two square numbers is always odd.

....

b) There are no numbers between 50 and 100 that are multiples of both 13 and 3.

....

c) Squaring a multiple of 4 will always give a number ending in 4 or 6.

....

You could also be asked to prove that something is or isn't a multiple of a number. For example, something is a multiple of 2 if you can write it as 2 × something else.

Q5 If $x = 3(y + 3) + 2(y - 2)$, show that x is a multiple of 5 for any whole number value of y.

....

....

Q6 Given that $15n + 24 = 3(m + 2n)$ show that m cannot be a multiple of 3 for any whole number n.

....

....

Coordinates and Midpoints

Remember — 1) x comes before y

2) x goes a-cross (get it) the page. (Ah, the old ones are the best...)

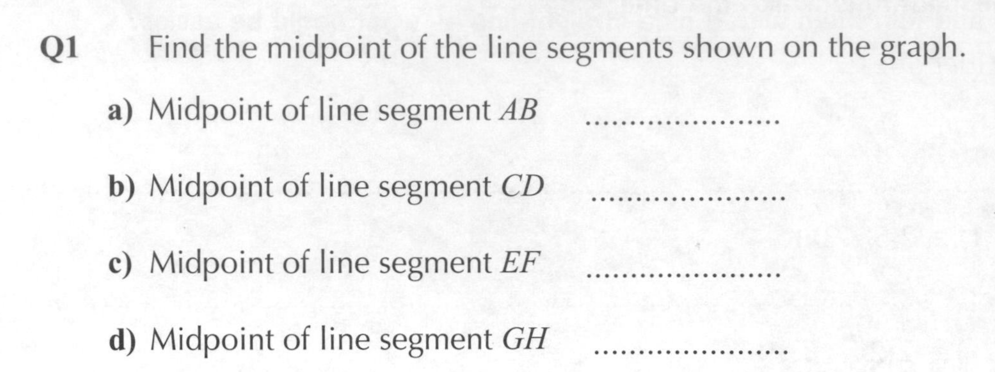

Q1 Find the midpoint of the line segments shown on the graph.

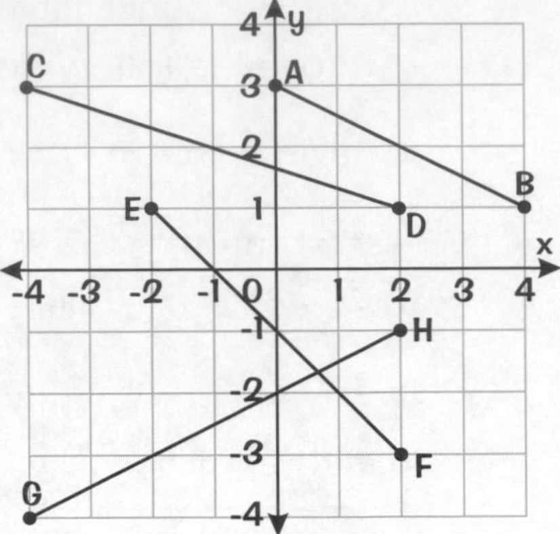

a) Midpoint of line segment AB ....

b) Midpoint of line segment CD ....

c) Midpoint of line segment EF ....

d) Midpoint of line segment GH ....

Q2 Find the midpoint of the line segments AB, where A and B have coordinates:

a) $A(2, 3)$ $B(4, 5)$ ....

b) $A(1, 8)$ $B(9, 2)$ ....

c) $A(0, 11)$ $B(12, 11)$ ....

d) $A(3, 15)$ $B(13, 3)$ ....

e) $A(6, 6)$ $B(-2, -2)$ ....

f) $A(15, -9)$ $B(-3, 3)$ ....

Q3 The isosceles triangle ABC shown below has two equal sides of length 6 units. Find the coordinates of point A.

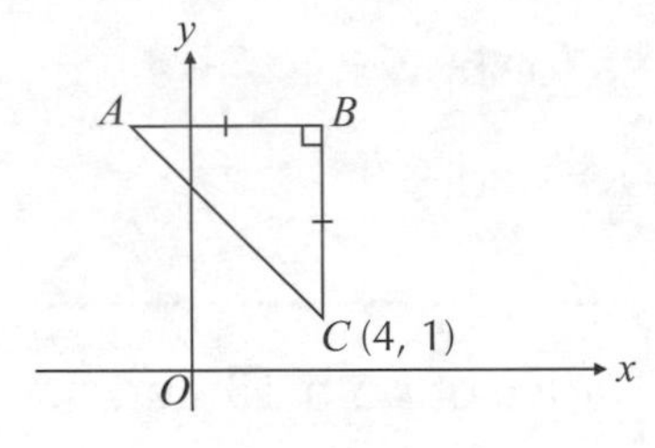

....

Q4 Find the coordinates of the missing endpoint of each line segment below.

a) Line segment PQ, with midpoint (3, 5.5), where P has coordinates (1, 5). ....

b) Line segment AB, with midpoint (3.5, 1.5), where A has coordinates (3, 3). ....

c) Line segment RS, with midpoint (-4, 5), where S has coordinates (7, -8). ....

Q5 Anna is designing the plan of a kitchen on a computer. The coordinates of the room's corners on screen are (0, 10), (220, 10), (0, 260), (220, 260). The ceiling light must be exactly in the centre of the room. What will the coordinates of the light be?

....

Drawing Straight-Line Graphs

The <u>very first thing</u> you've got to do is draw some axes, then use the equation to work out a <u>table of values</u>. All you've got to do then is plot the points from your table and join them with a nice straight line — what could be easier?

Q1 On the grid shown, draw axes with x from 0 to 8 and y from 0 to 14.

Q2 a) Complete the table of values for $y = x + 2$.

x	0	1	2	3	4	5	6
y	2			5			

b) Use your table of values to draw the graph of $y = x + 2$ on the grid opposite.

Q3 a) Fill in the table of values for $y + x = 8$.

x	0	1	2	3	4	5	6
y							

b) Draw the graph of $y + x = 8$ on the grid opposite.

Q4 A rough way of changing a temperature from Celsius to Fahrenheit is to 'double it and add 30' ($y = 2x + 30$).

a) Fill in the table below by changing the temperatures in Celsius (x) to Fahrenheit (y).

x	0	5	10	15	20	25	30
y	30				70		

b) Use your table of values to draw a temperature conversion graph. Draw and label the axes with x from 0 to 30, and y from 0 to 100.

Q5 a) Draw and label a pair of axes, with x from -3 to 3 and y from -3 to 15.

b) Complete the table of values for $y - 3x = 6$.

x	-3	-2	-1	0	1	2	3
y							

c) Use your table of values to draw a graph of $y - 3x = 6$ on your axes.

Straight-Line Graphs — Gradients

Now it's time to find the gradient of a line and look a bit more closely at the equations of straight lines. Remember — you find the gradient by working out $\frac{\text{change in y}}{\text{change in x}}$, simple!

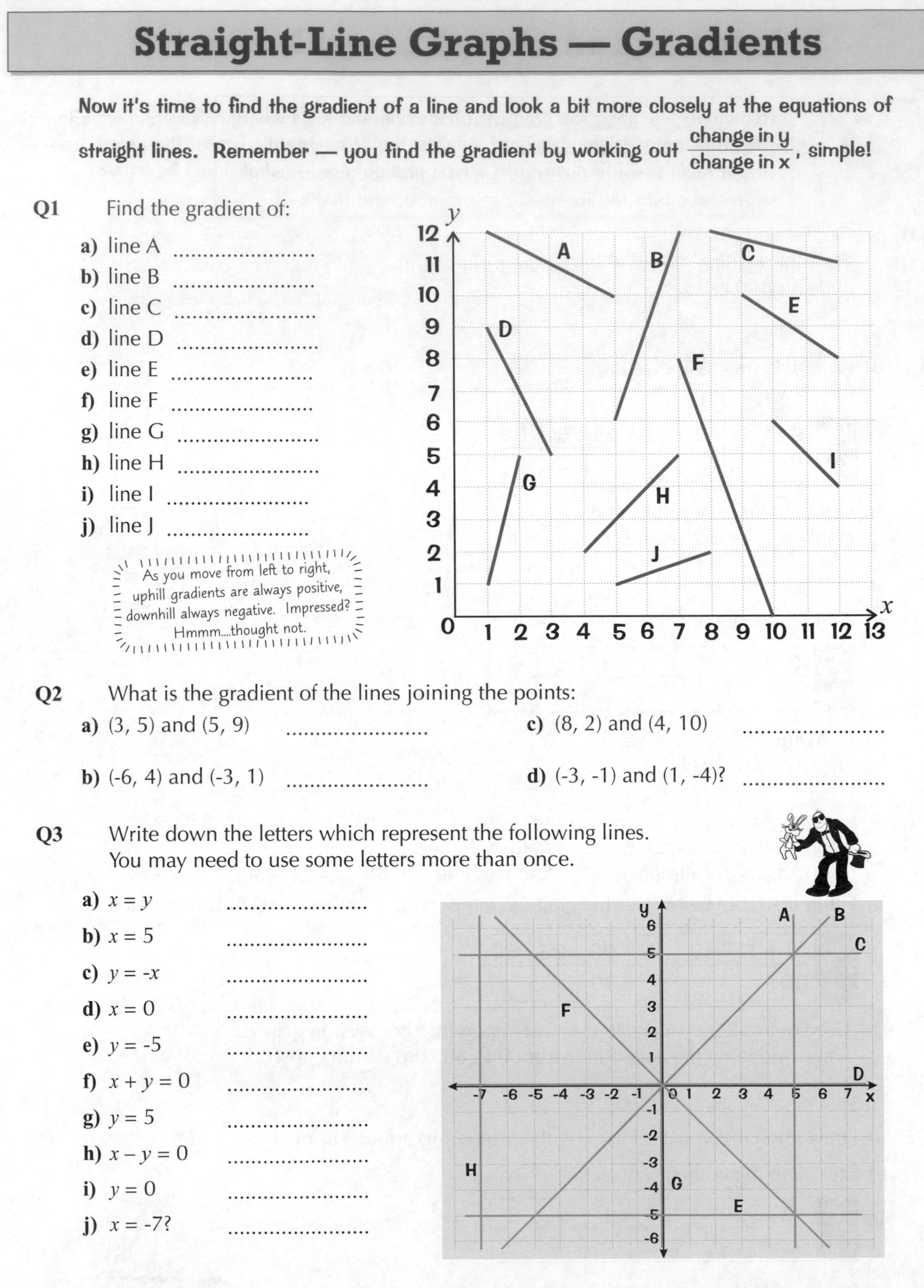

Q1 Find the gradient of:

a) line A ....
b) line B ....
c) line C ....
d) line D ....
e) line E ....
f) line F ....
g) line G ....
h) line H ....
i) line I ....
j) line J ....

As you move from left to right, uphill gradients are always positive, downhill always negative. Impressed? Hmmm....thought not.

Q2 What is the gradient of the lines joining the points:

a) (3, 5) and (5, 9) ....
b) (-6, 4) and (-3, 1) ....
c) (8, 2) and (4, 10) ....
d) (-3, -1) and (1, -4)? ....

Q3 Write down the letters which represent the following lines.
You may need to use some letters more than once.

a) $x = y$ ....
b) $x = 5$ ....
c) $y = -x$ ....
d) $x = 0$ ....
e) $y = -5$ ....
f) $x + y = 0$ ....
g) $y = 5$ ....
h) $x - y = 0$ ....
i) $y = 0$ ....
j) $x = -7$? ....

Q4 Which of the following equations are straight lines?

a) $y = 4x + 6$ **b)** $y = 6x - 3$ **c)** $3y = x^2 + 2$ **d)** $2y = 4 - 5x$

Straight-Line Graphs — y = mx + c

Remember — 'm' is the gradient of the line and 'c' is where the graph crosses the y-axis (it's called the 'y-intercept').
If you need to write the equation of a line, pop your values of m and c into the formula y = mx + c, and that's it!

Q1 The following are equations of straight-line graphs.
Without plotting the graphs, state the gradient of each graph and the y-intercept.

a) $y = 4x + 2$ ....
e) $y = x$ ....

b) $y = 5x - 1$ ....
f) $y = 3 - x$ ....

c) $y = 6x$ ....
g) $y + 2x = 10$ ....

d) $y = 5 + 2x$ ....
h) $2y = x + 4$ ....

Q2 Find the values for m and c if the straight-line graph $y = mx + c$ has a gradient of 3 and passes through (0, 8).

....

Q3 For each of these lines, find the gradient, m, and the y-intercept, c.
Hence write down the equation of the line.

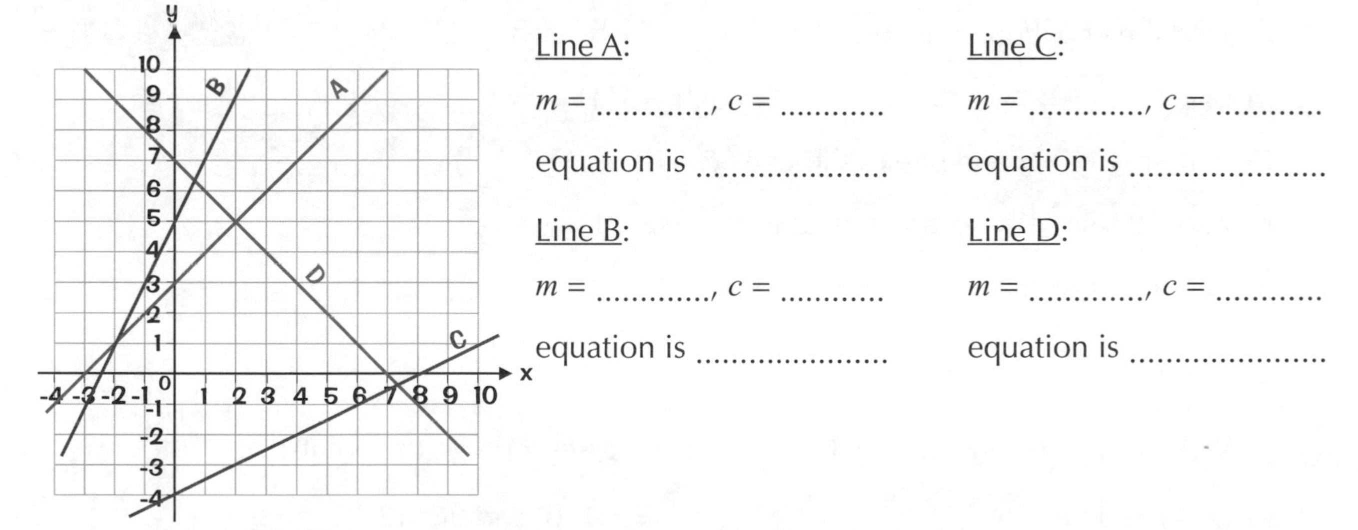

Line A:
$m =$ ...., $c =$ ....
equation is ....

Line B:
$m =$ ...., $c =$ ....
equation is ....

Line C:
$m =$ ...., $c =$ ....
equation is ....

Line D:
$m =$ ...., $c =$ ....
equation is ....

Q4 In each of the following, circle the points that lie on the given line.

a) $y = 3x - 1$ (7, 20) (6, 15) (5, 14)

b) $y = 2x + 2$ (-1, 0) (2, 8) (6, 12)

c) $y = -7x + 9$ (-4, 37) (-3, 25) (2, -5)

Try substituting the x and y coordinates into the equation to see if the equation is true at that point.

Using y = mx + c

There are two main things to practise on this page — using y = mx + c as a formula and working with parallel lines. Remember — when lines are parallel, their gradients are the same.

Q1 Put a circle round the pairs of straight lines below which are parallel.

You only need to compare the gradients to answer Q1.

a) $y = 3x + 4, \quad y = 3x - 8$

b) $y = 2x - 17, \quad y - 2x = 17$

c) $y = x - 4, \quad y = -2x - 4$

d) $3y + 3x = 9, \quad y = x + \frac{3}{2}$

e) $2x = y + 8, \quad 2y = x + 8$

f) $3 - x = 3y, \quad y - 3x = 5$

g) $4y + 8x = 0, \quad y = 22 - 2x$

h) $y = 2x - 8, \quad 8y + 8x = 9$

Q2 Line A has the equation $y = 3x - 4$.
Line B is parallel to line A and passes through (8, 25).

a) Find the gradient of line B. ....

b) Find the equation for line B. ....

Q3 Find the equation of the straight line which passes through:

a) (3, 7) and has a gradient of 1 ....

b) (2, 8) and has a gradient of 3 ....

c) (4, -4) and has a gradient of -1 ....

d) (-1, 7) and has a gradient of -3 ....

Q4 Find the equation of the line which is:

a) parallel to $y = 2x - 1$ and passes through (4, 11) ....

b) parallel to $y = x$ and passes through (-2, -10) ....

c) parallel to $y = 1 - 5x$ and passes through (2, 6) ....

d) parallel to $y = 8x - 4$ and passes through (3, 12) ....

Find the gradient of the line, then find the y-intercept.

Q5 Write down the equation of the line which passes through the points:

a) (2, 2) and (5, 5) ....

b) (1, 3) and (4, 12) ....

c) (1, 0) and (5, -12) ....

d) (-5, 6) and (-1, -2) ....

Q6 Find the values of *a*-*c* if:

a) the point (4, *a*) is on the line $y = 2x - 1$ ....

b) the point (-3, *b*) is on the line $y = -3x$? ....

c) the point (*c*, 13) is on the line $y = 3x + 1$ ....

Quadratic Graphs

- **If an expression has an x^2 term in it (and no higher powers of x), it's quadratic. The graphs you get from quadratic expressions are always curves shaped like a bucket. (Funny shaped bucket, but you see what I mean.)**

Q1 Use the axes on the right to sketch the graph of $y = x^2$.

Label where the graph meets the axes.

Never use a ruler to sketch or draw a quadratic graph — it should always be a smooth curve.

Q2 **a)** Complete this table of values for the graph $y = 2x^2$.

x	-4	-3	-2	-1	0	1	2	3	4
x^2	16	9					4		
$y=2x^2$	32	18					8		

b) Plot the graph $y = 2x^2$, using axes with x from -4 to 4 and y from 0 to 32. Label your graph.

Q3 **a)** Complete this table of values for the graph $y = x^2 - 4x + 1$.

x	-2	-1	0	1	2	3	4
x^2	4	1				9	
$-4x$	8					-12	
1	1	1				1	
$y=x^2-4x+1$	13	6				-2	

b) Plot the graph $y = x^2 - 4x + 1$, using axes with x from -2 to 4 and y from -3 to 13.

c) Draw and label the line of symmetry.

If the x^2 term has a minus sign in front of it, the curve will be upside down.

Q4 **a)** Complete this table of values for the graph $y = 3 - x^2$.

x	-4	-3	-2	-1	0	1	2	3	4
3	3	3	3	3	3	3	3	3	3
$-x^2$	-16						-4		
$y=3-x^2$	-13						-1		

b) Draw the graph $y = 3 - x^2$ for x from -4 to 4.

Q5 The curve on the right has the equation $y = 2x^2 + 6x + c$.

a) Find the value of c.

....

b) Write down the coordinates of the turning point.

....

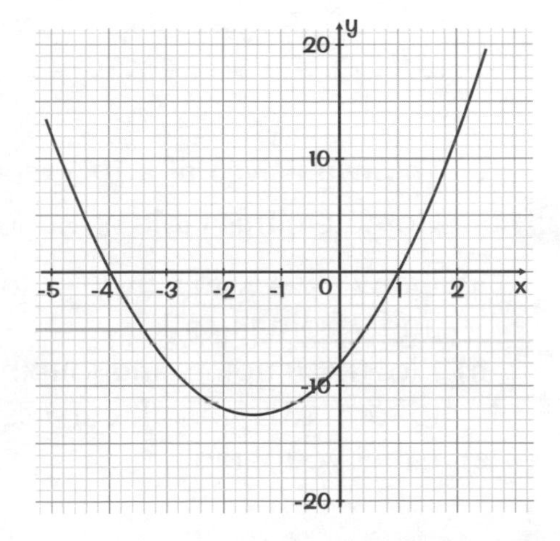

Harder Graphs

You should know the basic shapes for quadratic, cubic and reciprocal (i.e. 1/x) graphs. They might look complicated at first, but you'll soon be able to spot them a mile away.

Q1 Here are some equations. Match each equation to one of the curves below.

a) $y = -x^2$ ....
b) $y = \frac{1}{x}$ ....
c) $y = -x^2 - 3$ ....
d) $y = -\frac{1}{2}x^3 + 2$ ....
e) $y = -x^3 + 3$ ....
f) $y = x^3$ ....

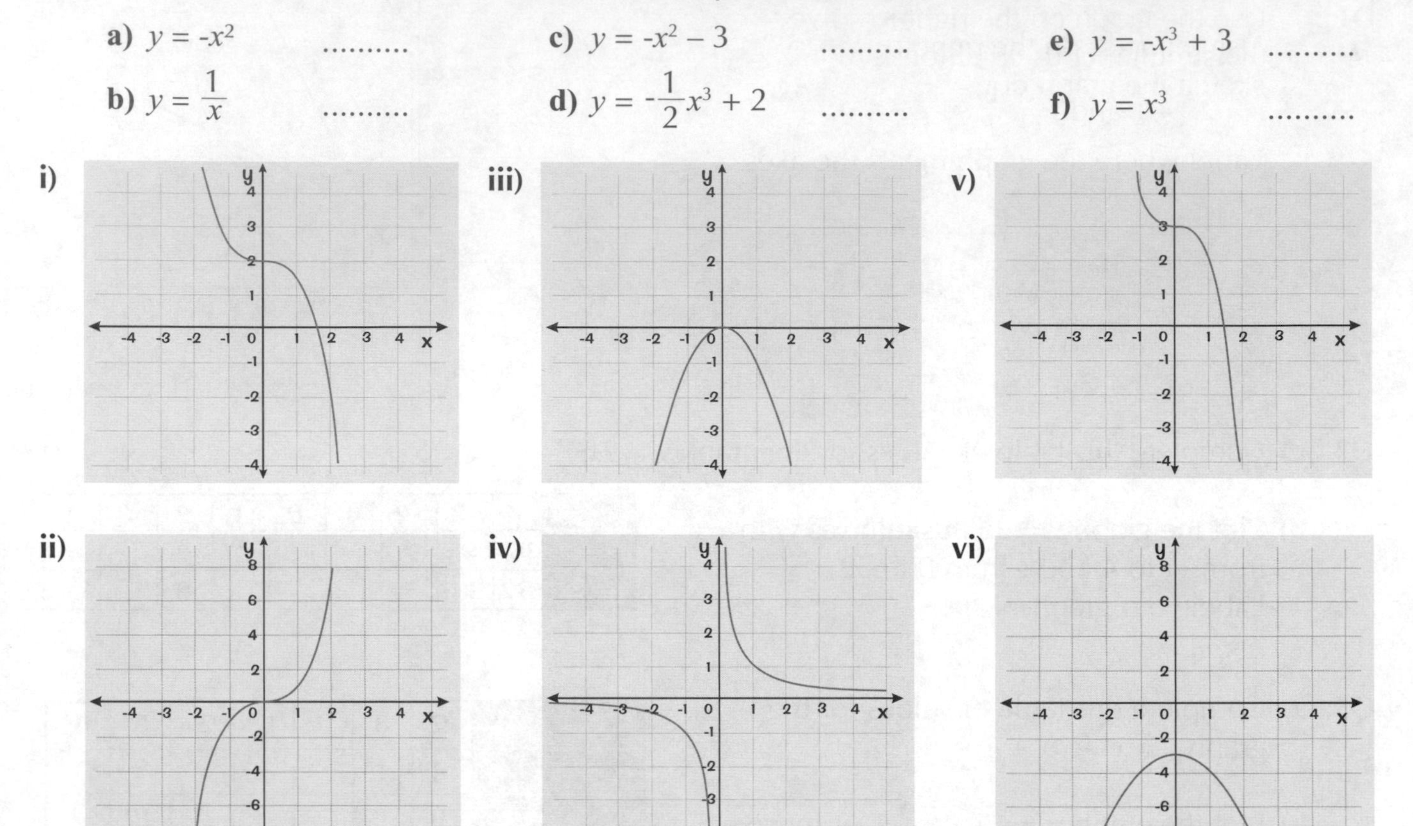

Q2 Complete this table of values for $y = x^3 + 4$:

x	-3	-2	-1	0	1	2	3
x^3							
+4							
y							

Draw the graph of $y = x^3 + 4$.

Q3 Complete this table of values for $y = -x^3 + 3$:

x	-3	-2	-1	0	1	2	3
$-x^3$							
+3							
y							

Draw the graph of $y = -x^3 + 3$.

The two halves of a 1/x graph <u>never touch</u> and are <u>symmetrical</u> around y = x and y = -x.

Q4 Complete this table of values for the graph $y = 1/x$.

a) Draw axes with x from -4 to 4 and y from -1 to 1.
b) Plot these points and join them to form two smooth curves.
c) Label your graph.

x	-4	-3	-2	-1	0	1	2	3	4
y=1/x		-0.33			n/a				0.25

The O's just there to fool you — any equation with A/x in it never makes a graph that passes through $x = 0$.

Solving Equations Using Graphs

This just in: Graphs aren't just pretty things to decorate your walls with — you can also use 'em to solve equations. They're especially useful for finding solutions to quadratics.

Q1 Use the graph on the right to solve the simultaneous equations $y = 2x - 1$ and $y = \frac{1}{2}x + 2$.

....

Psssst — 'solving the simultaneous equations' just means finding the x- and y-value of the point where the graphs cross.

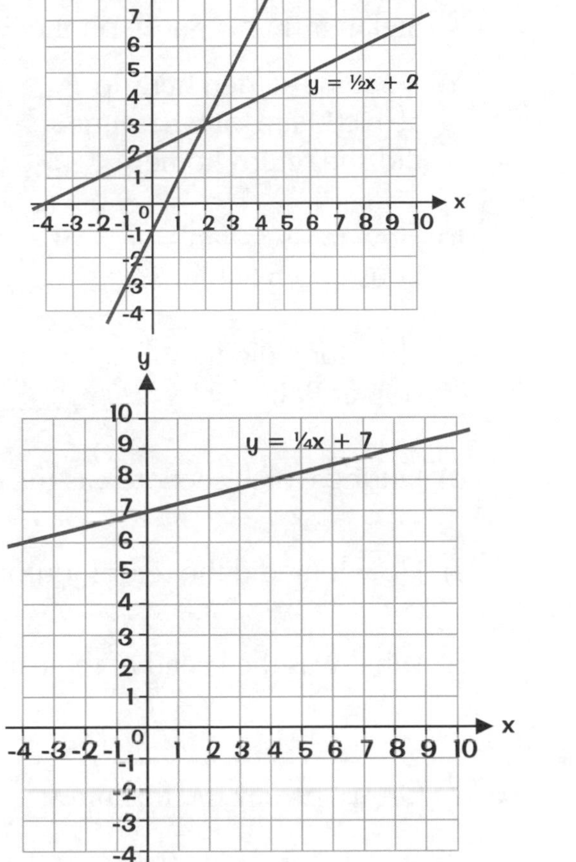

Q2 The graph of $y = \frac{1}{4}x + 7$ is shown on the right. Use the graph to find the solution to $2x = \frac{1}{4}x + 7$.

....

Q3 Solve the following simultaneous equations by drawing graphs. Use axes with x from 0 to 6.

a) $y = x$
$y = 9 - 2x$

....

b) $y = 2x + 1$
$2y = 8 + x$

....

c) $y = 4 - 2x$
$x + y = 3$

....

d) $y = 3 - x$
$3x + y = 5$

....

Q4 The curve on the right has the equation $y = x^2 + 2x - 8$.

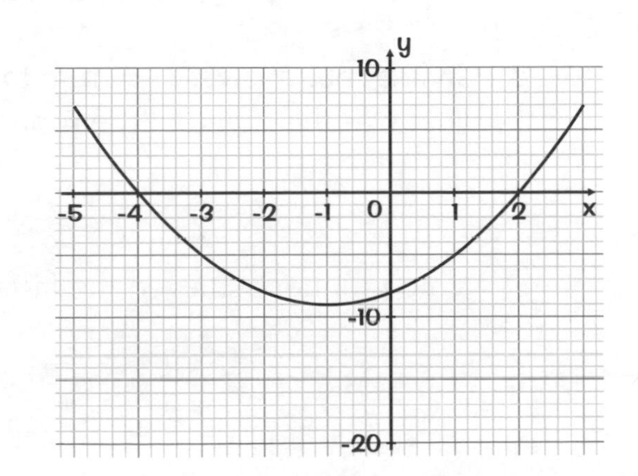

a) Use the curve to find both roots of $x^2 + 2x - 8 = 0$.

....

b) Use the curve to find the solutions of the equation $x^2 + 2x - 8 = -5$.

....

Distance-Time Graphs

Ahh, distance-time graphs — one of the old classics of GCSE Maths. They'll still be around long after everything is rubble and the insects have taken over the earth. Ahem, sorry — that's for a whole different book. The thing to remember with these graphs is that the <u>gradient</u> is the <u>speed</u>.

Q1 The graph shows Sabine's car journey from her house to Ralf's house and back, picking up Stefan from his house on the way.

You can work out where the houses are by looking for the flat parts of the graph — the bits where Sabine stops.

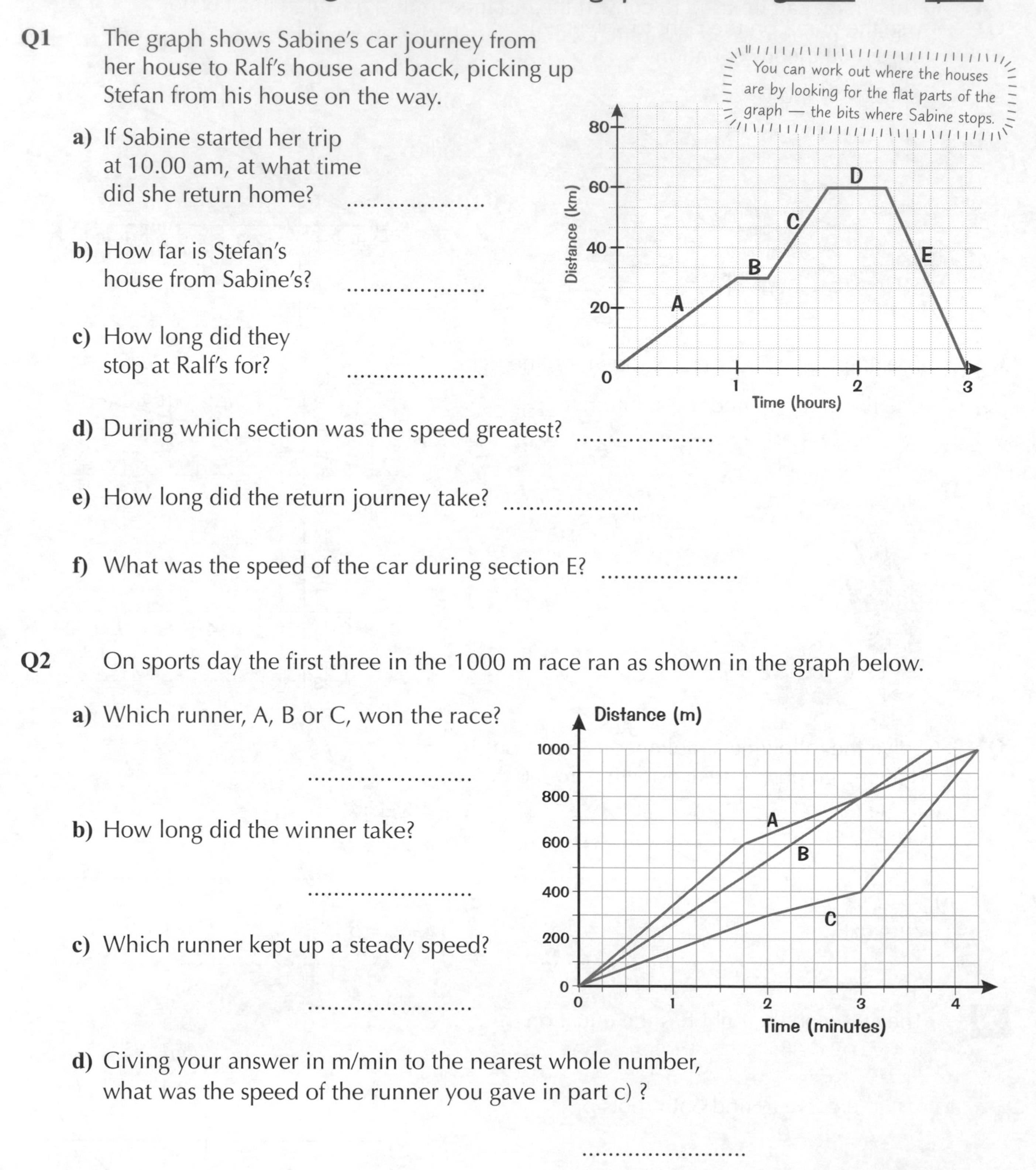

a) If Sabine started her trip at 10.00 am, at what time did she return home? ....

b) How far is Stefan's house from Sabine's? ....

c) How long did they stop at Ralf's for? ....

d) During which section was the speed greatest? ....

e) How long did the return journey take? ....

f) What was the speed of the car during section E? ....

Q2 On sports day the first three in the 1000 m race ran as shown in the graph below.

a) Which runner, A, B or C, won the race?

....

b) How long did the winner take?

....

c) Which runner kept up a steady speed?

....

d) Giving your answer in m/min to the nearest whole number, what was the speed of the runner you gave in part c) ?

....

e) Which runner achieved the fastest speed at any point in the race?

....

Real-Life Graphs

In that 'real life' that I've heard so much about, graphs are a handy way of converting measurements from one unit to another, or calculating how much something will cost...

Q1 This graph can be used to convert the distance (miles) travelled in a taxi to the fare payable (£). How much will the fare be if you travel:

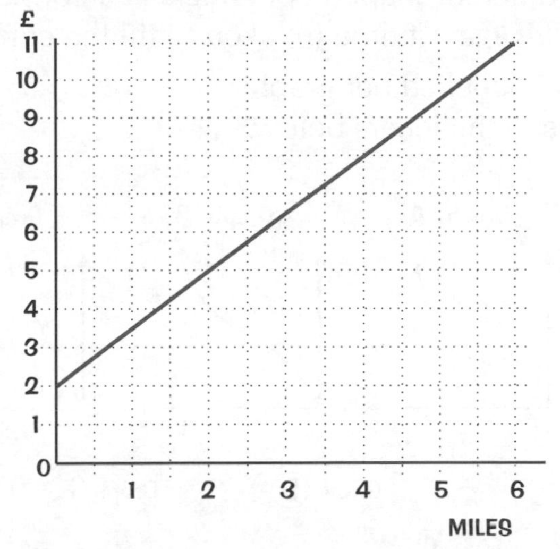

a) 2 miles ....

b) 5 miles ....

c) 10 miles ....

d) Mike lives 4.5 miles away from the cinema. Is £16 enough money for Mike to get a taxi to the cinema and then home again after watching a film?

....

Q2 80 km is roughly equal to 50 miles. Use this information to draw a conversion graph on the grid. Use the graph to estimate the number of miles equal to:

a) 20 km ....

b) 70 km ....

c) 90 km ....

Q3 Use the same graph to find the number of km equal to:

a) 40 miles ....

b) 10 miles ....

c) 30 miles ....

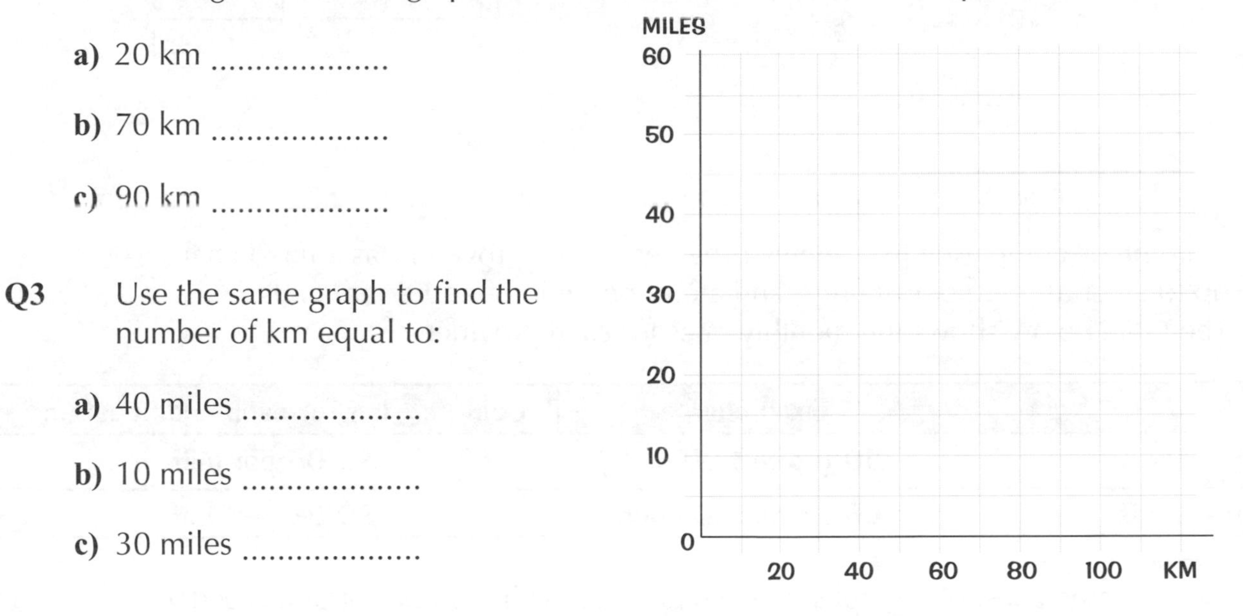

Q4 The graph below shows the cost of hiring a holiday cottage for one week.

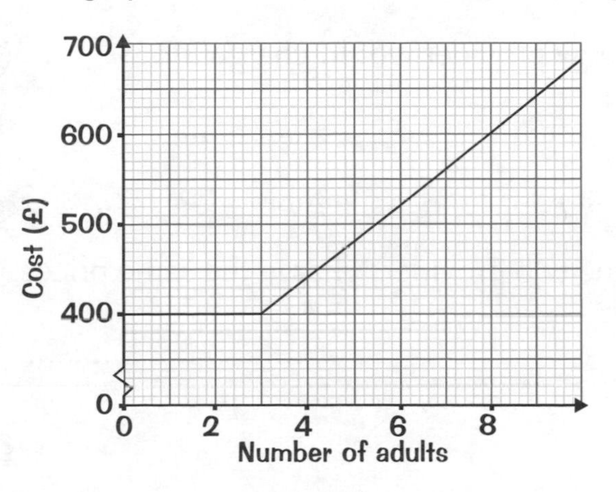

a) How many adults are included in the basic weekly cost?

....

b) Estimate the cost per adult for additional adults.

....

Real-Life Graphs

A couple more examples of how graphs can be used outside of the maths classroom...

Q5 In her science lesson, Ellie pours water into different shaped containers at a constant rate, then plots graphs of the depth of water (d) against time (t) taken to fill the container.
At the end of the lesson she realises she hasn't labelled her graphs.
Which graph matches each container? Write in the letters below.

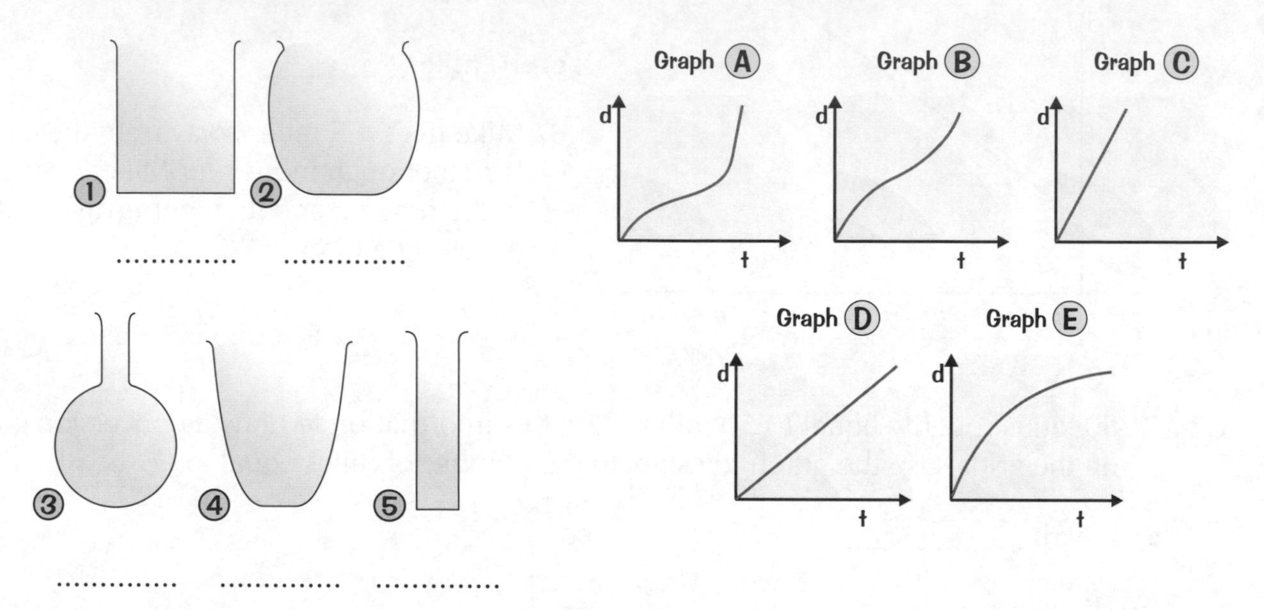

Q6 Dominic is comparing two energy providers. Each provider has a fixed charge up to a certain number of units, and a cost per unit for additional units.
The table below shows the monthly rates for each provider.

	Fixed charge	Cost per unit for each additional unit
Provider A	**£10 (up to 200 units)**	**£0.05 per unit**
Provider B	**£5 (up to 150 units)**	**£0.06 per unit**

a) On the same axes, plot graphs for each provider with 'number of units used' on the x-axis and 'monthly cost' on the y-axis.

b) Use your graph to find the price each energy provider would charge for:

i) 250 units per month A: .... B: ....

ii) 460 units per month A: .... B: ....

c) For what number of units per month do the two providers charge the same price?

....

Ratios

Ratios compare quantities of the same kind — so if the units aren't mentioned, they've got to be the same in each bit of the ratio.

Q1 Write each of these ratios in its simplest form. The first one is done for you.

a) 4:6 **b)** 15:21 **c)** 72:45 **d)** 24 cm:36 cm

2:3 ...... : .... : .... : ......

e) 350 g:2 kg **f)** 42p:£1.36 **g)** 2.9:8.7 **h)** $\frac{3}{5}:\frac{9}{10}$

...... : .... : .... : .... : ......

Watch out for ones like e) and f) — you need to make the units the same first.

Q2 Reduce each of the following ratios to the form 1:n. The first one is done for you.

a) 2:18 **b)** 5:35 **c)** 4:26 **d)** 12 cm:4.5 m

1:9 .... : .... : .... : ....

Q3 Reduce each of the following ratios to the form n:1. The first one is done for you.

a) 12:4 **b)** 19:5 **c)** 17 cm:0.02 m **d)** 8 m:40 cm

3:1 .... : .... : .... : ....

Q4 To make grey paint, black and white paint are mixed in the ratio 5:3.
How much black paint would be needed with:

a) 6 litres of white? ....

b) 12 litres of white? ....

c) 21 litres of white? ....

Q5 To make salad dressing you mix vinegar and olive oil in the ratio 2:5.
How much olive oil is needed with:

a) 10 ml of vinegar? .... **b)** 30 ml of vinegar? ....

c) 42 ml of vinegar? ....

Q6 The ratio of men to women at a football match was 11:4.
How many men were there if there were:

a) 2000 women? .... **b)** 8460 women? ....

Ratios

Q7 For each of the following ratios, complete the statements that describe them by filling each gap with a fraction:

a) Cats and dogs in the ratio 2 : 5.

i) There are ...... as many cats as dogs.

ii) There are ...... times as many dogs as cats.

b) Sprouts and peas in the ratio 3 : 4.

i) There are ...... as many sprouts as peas.

ii) There are ...... times as many peas as sprouts.

Q8 Alex's sock drawer contains plain, spotted and striped socks in the ratio 5 : 3 : 8. What fraction of Alex's socks are striped? Give your answer in its simplest form.

....

Q9 Heather has red and green pens in her stationery cupboard. The ratio of green pens to the total number of pens is 4 : 11.

a) What fraction of Heather's pens are green? ....

b) What is the ratio of green pens to red pens? ....

c) Heather has 56 red pens. How many green pens does she have? ....

If a question gives you an amount to split into a ratio, a great way to check your answer works is to add up the individual quantities — they should add up to the original amount.

Q10 Sarah works as a waitress. Each week, she splits her wage into spending money and savings in the ratio 7 : 3.

a) One week, Sarah earns £130. How much should she put in her savings that week? ....

b) The next week, Sarah put £42 into her savings. How much did she earn in total that week? ....

Q11 Divide each of the following quantities in the given ratio.

a) 100 g in the ratio 1 : 4 = ...

b) 500 m in the ratio 2 : 3 = ...

c) £12 000 in the ratio 1 : 2 = ...

d) 6.3 kg in the ratio 3 : 4 = ...

e) £8.10 in the ratio 4 : 5 = ...

Ratios

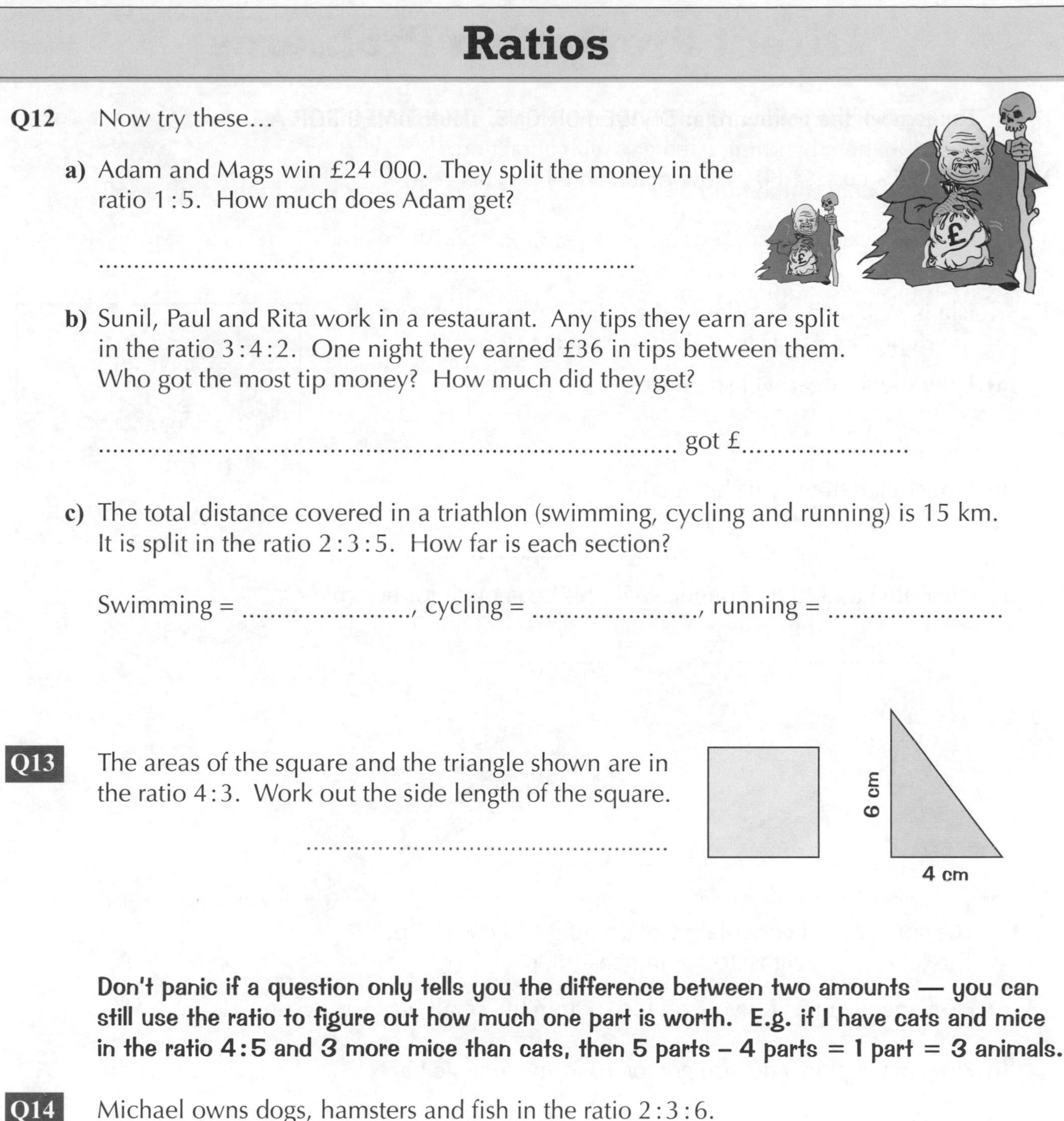

Q12 Now try these...

a) Adam and Mags win £24 000. They split the money in the ratio 1 : 5. How much does Adam get?

....

b) Sunil, Paul and Rita work in a restaurant. Any tips they earn are split in the ratio 3 : 4 : 2. One night they earned £36 in tips between them. Who got the most tip money? How much did they get?

.... got £......................

c) The total distance covered in a triathlon (swimming, cycling and running) is 15 km. It is split in the ratio 2 : 3 : 5. How far is each section?

Swimming = ..., cycling = ..., running = ...

Q13 The areas of the square and the triangle shown are in the ratio 4 : 3. Work out the side length of the square.

....

Don't panic if a question only tells you the difference between two amounts — you can still use the ratio to figure out how much one part is worth. E.g. if I have cats and mice in the ratio 4 : 5 and 3 more mice than cats, then 5 parts – 4 parts = 1 part = 3 animals.

Q14 Michael owns dogs, hamsters and fish in the ratio 2 : 3 : 6.
He has 2 more hamsters than dogs. How many fish does Michael have?

....

Q15 Fay, Ben and Peggy count the money they each have in their pocket.
Ben has 4 times more than Fay. Peggy has 6 times more than Fay.

a) Write down the ratio of their amounts in its simplest form.

....

b) Ben has £9 more than Fay. How much does Peggy have?

....

Direct Proportion Problems

Remember the golden rule: DIVIDE FOR ONE, THEN TIMES FOR ALL

Q1 7 pencils cost £2.80. How much will 4 pencils cost?

....

Q2 Isla is making a chocolate cake using the recipe shown on the right. She wants to make the cake for 10 people.

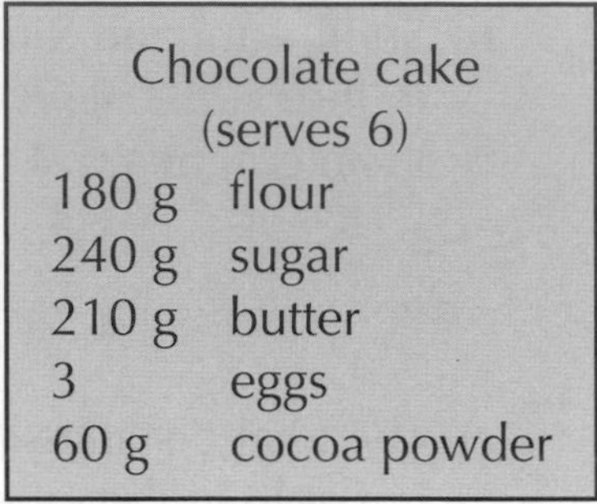

Chocolate cake
(serves 6)

180 g	flour
240 g	sugar
210 g	butter
3	eggs
60 g	cocoa powder

a) How much sugar will she need?

....

b) How much flour will she need?

....

c) Isla only has 320 g of butter. Will this be enough for her cake?

....

Q3

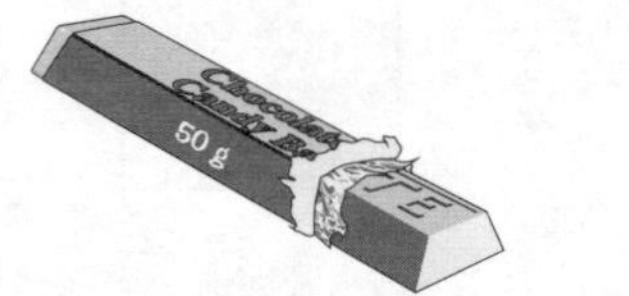

The small bar of chocolate weighs 50 g and costs 32p.
The large bar weighs 200 g and costs 80p.

a) How many grams do you get for 1p from the small bar? ....

b) How many grams do you get for 1p from the large bar? ....

c) Which bar gives you more for your money? ....

Q4 Which of these boxes of eggs is better value for money?

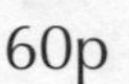

60p

£1.30

....

Q5 A shop sells bags of crisps in three different sizes. A 30 g bag costs 65p, a 160 g bag costs £2.55 and a 250 g bag costs £3.49. Which of the bags represents the best value for money?

....

Direct Proportion Problems

Q6 Alice takes the bus whenever she goes shopping in town. The bus ticket costs the same each time. Last month, Alice went shopping 3 times and the bus tickets cost her £6.75 in total. This month, Alice spent £11.25 on bus tickets for her shopping trips. How many times did she go shopping this month?

....

Q7 At a badminton club, $\frac{4}{9}$ of the members are men. 15 of the members are women. How many of the members are men?

Work out what fraction of the members are women first.

....

Q8 Angela pays £408 for 3 months' worth of school dinners for her 4 children. What would the cost of 5 months' worth of school dinners be for 3 children?

Work out the cost for 1 child for 1 month first.

....

Q9 Ian drinks 4 cups of tea every day, using one teabag in each cup. He has 50 ml of milk in each cup of tea. Ian pays £3 for a box of 140 teabags and 55 p for a 500 ml bottle of milk. What is the total cost of the cups of tea Ian drinks in one week?

....

Q10 y is directly proportional to x. When $x = 3$, $y = 12$.

a) Write an equation of the form $y = Ax$ to represent this proportion.

....

b) Sketch the graph of this proportion, marking two points on the line.

Q11 A recipe for a pasta sauce contains tomatoes and peppers in the ratio 5 : 2. Sketch a graph showing the number of tomatoes against the number of peppers in the sauce. Mark two points on the line.

Inverse Proportion Problems

With inverse proportion, one value decreases as the other increases.

Q1 Circle each of the equations below that show that y is inversely proportional to x.

$y = \frac{1}{x}$ $x = \frac{y}{5}$ $\frac{1}{x} = \frac{y}{3}$ $y = 7x$ $x = \frac{8}{y}$ $x = \frac{1}{y} + 4$

Q2 Given that a is inversely proportional to b, complete the statements below:

a) If a is multiplied by 3 then b is ....

b) If a is divided by 5 then b is ....

Q3 y is inversely proportional to x. When $y = 2$, $x = 6$.

a) Find the value of y when $x = 4$. ....

b) Sketch the graph of this proportion in the space on the right. Mark two points on the graph.

The golden rule for inverse proportion is the exact opposite of the rule for direct proportion: TIMES FOR ONE, THEN DIVIDE FOR ALL.

Q4 It takes two people three hours to wrap 90 presents. Working at the same rate, how long would it take three people to wrap 90 presents? ....

Q5 4 people can paint 12 bikes in 9 hours. How much longer would it take 3 people to paint the same number of bikes if they worked at the same rate?

....

Q6 Mona is planning a cycle ride. The time it will take Mona to complete the route is inversely proportional to the speed at which she travels. She knows that if she travels at 15 mph, it will take her 2 hours to complete the route.

Plot a graph showing the relationship between Mona's speed and the time it will take her to cycle the route.

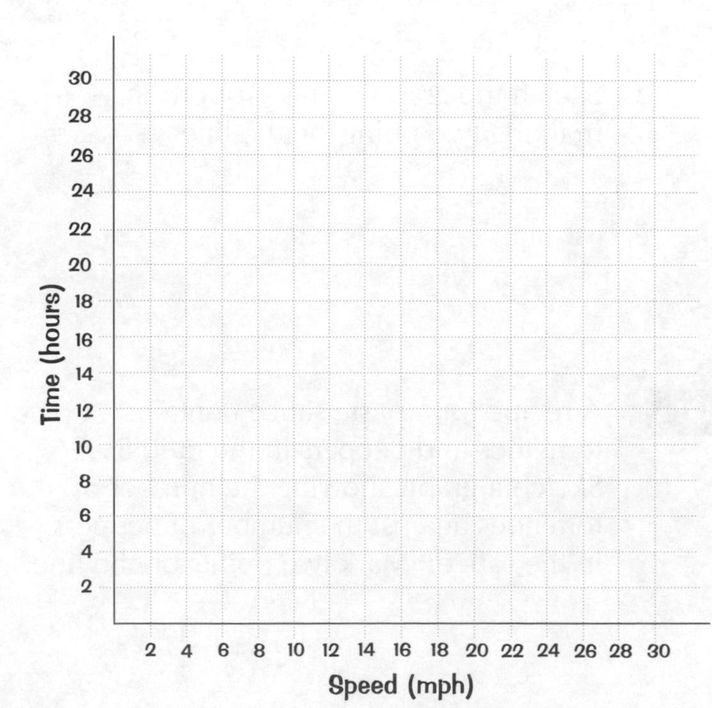

Percentages

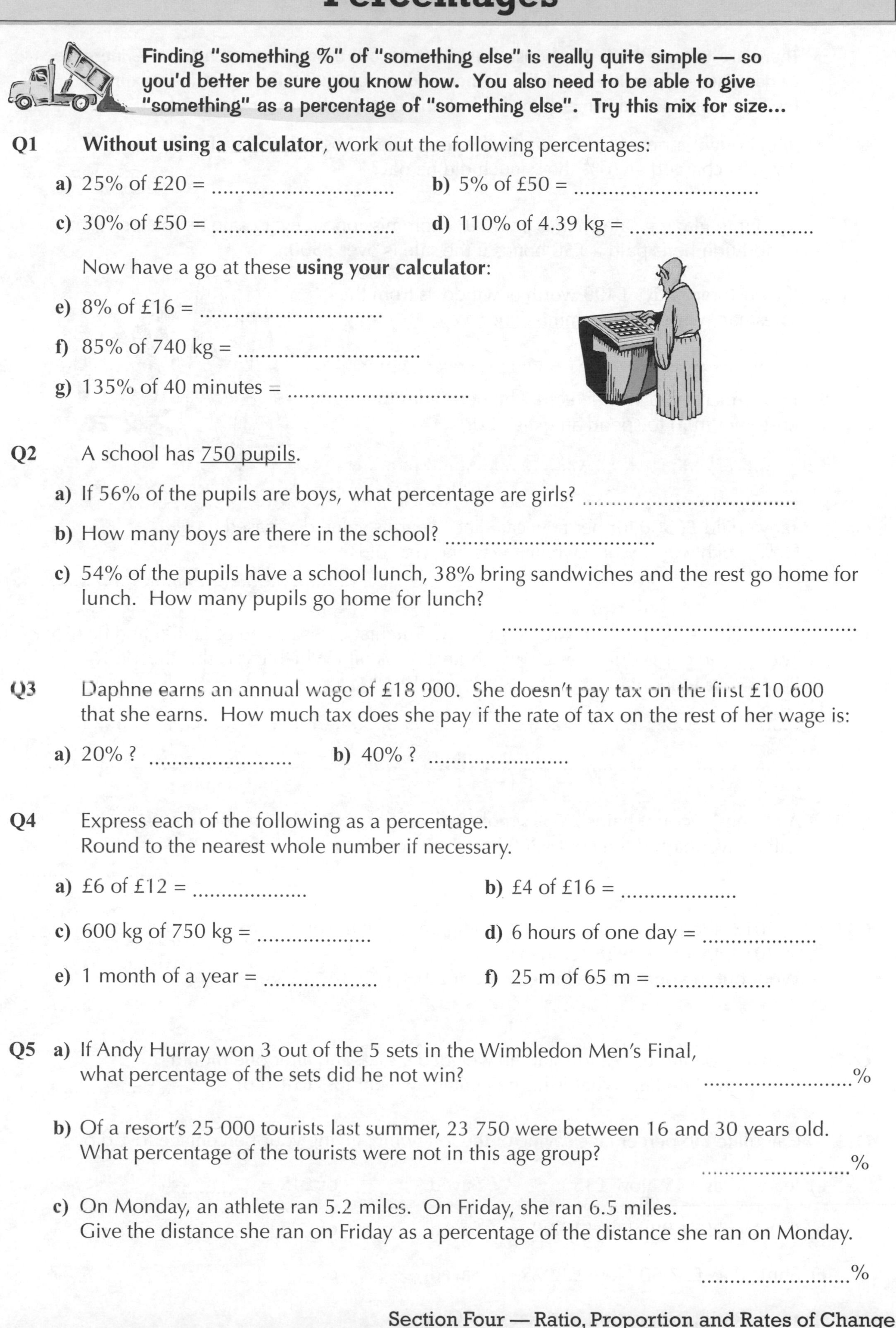

Finding "something %" of "something else" is really quite simple — so you'd better be sure you know how. You also need to be able to give "something" as a percentage of "something else". Try this mix for size...

Q1 **Without using a calculator**, work out the following percentages:

a) 25% of £20 = ...

b) 5% of £50 = ...

c) 30% of £50 = ...

d) 110% of 4.39 kg = ...

Now have a go at these **using your calculator**:

e) 8% of £16 = ...

f) 85% of 740 kg = ...

g) 135% of 40 minutes = ...

Q2 A school has 750 pupils.

a) If 56% of the pupils are boys, what percentage are girls? ....

b) How many boys are there in the school? ....

c) 54% of the pupils have a school lunch, 38% bring sandwiches and the rest go home for lunch. How many pupils go home for lunch?

....

Q3 Daphne earns an annual wage of £18 900. She doesn't pay tax on the first £10 600 that she earns. How much tax does she pay if the rate of tax on the rest of her wage is:

a) 20% ? ....

b) 40% ? ....

Q4 Express each of the following as a percentage.
Round to the nearest whole number if necessary.

a) £6 of £12 = ...

b) £4 of £16 = ...

c) 600 kg of 750 kg = ...

d) 6 hours of one day = ...

e) 1 month of a year = ...

f) 25 m of 65 m = ...

Q5 **a)** If Andy Hurray won 3 out of the 5 sets in the Wimbledon Men's Final, what percentage of the sets did he not win?

.........................%

b) Of a resort's 25 000 tourists last summer, 23 750 were between 16 and 30 years old. What percentage of the tourists were not in this age group?

.........................%

c) On Monday, an athlete ran 5.2 miles. On Friday, she ran 6.5 miles.
Give the distance she ran on Friday as a percentage of the distance she ran on Monday.

.........................%

Percentages

There are two methods to choose from to work out a new amount after a % increase or decrease — one is to find the % then add or subtract. The other is to express the % change as a decimal (a multiplier) and multiply the original value by it.

Q6 John bought a new PC. The tag in the shop said it cost £890 + VAT.
If VAT is charged at 20%, how much did he pay?

....

Q7 A double-glazing salesman is paid 10% commission on every sale he makes.
In addition he is paid a £50 bonus if the sale is over £500.

a) If a customer buys £499 worth of windows from the salesman, what is his commission?

....

b) How much extra will he earn if he persuades the customer in **a)** to spend an extra £20?

....

Q8 Tanya paid £6500 for her new car. Each year its value decreased by 8%.
How much was it worth when it was one year old?

....

Q9 Tim is choosing between two cars to buy. The first car is priced at £8495 and has 15% off.
The second car is priced at £8195 and has 12% off. Which car is the cheapest?
Show your working.

....

....

Q10 A savings account gains 2.5% simple interest per year. How much interest will a saver earn if they put £200 into their account for three years?

....

Q11 In 2013, there were 250 students at Highmoor Sixth Form.
In 2014, there were 280 students.
Work out the percentage increase from 2013 to 2014.

Percentage change = $\frac{\text{change}}{\text{original}} \times 100$

.... %

Q12 Jeff went on a diet. At the start he weighed 92 kg, and after one month he weighed 84 kg. What is his percentage weight loss to 1 d.p?

..........................%

Q13 Calculate the percentage saving of the following, giving your percentage to 1 d.p.

a) Jeans: Was £45 Now £35 Saved .... of £45 = ...%

b) CD: Was £14.99 Now £12.99 Saved .... of .... = ...%

c) Shirt: Was £27.50 Now £22.75 Saved .... of .... = ...%

Percentages

Finding the original value always looks a bit confusing at first. The bit most people get wrong is deciding whether the value given represents more or less than 100% of the original — so always check your answer makes sense.

Q14 In the new year sales Robin bought a tennis racket for £68.00. The original price had been reduced by 15%. What was the original price?

....

Q15 There are 360 people living in a certain village. The population of the village has grown by 20% over the past year. How many people lived in the village one year ago?

....

Q16 The council tells Mrs Clarke that the tree in her garden is too tall, and orders her to reduce its height by 15%. She cuts 75 cm off the top of the tree to meet the regulations. How tall was her tree originally? Give your answer in m.

75 cm is 15% of the original height.

.... m

Some percentage questions can be a bit weird and wonderful, so you'll need to get your thinking cap on. But don't panic — as long as you remember all your percentage skills, you'll be fine.

Q17 In 2012, Julie's salary was £23 000. In 2013, her salary was 3% lower than in 2012. In 2014, Julie was given a promotion, and her salary was 12% higher than in 2013. What is the percentage increase in Julie's salary between 2012 and 2014?

Do this step by step — work out her salary in 2013, then use that value to work out her 2014 salary.

....

Q18 Emily has her own business making and selling greetings cards.
Each card costs her 68p to make. One day, Emily has 300 cards to sell.
She sells 65% of them for 99p each, and the rest for 75p each.
How much profit did Emily make?

£ ....

Q19 At a running club, 30% of the members are male. 60% of the male members and 50% of the female members are training to run a marathon. **Without using a calculator**, work out what percentage of the members are training for a marathon.

....

Q20 The ratio of fiction to non-fiction books in a library is 7 : 3.
60% of the fiction books are hardback, and 20% of the non-fiction books are hardback.
What percentage of all the books in the library are hardback?

Start by turning the ratio into a percentage.

.... %

Compound Growth and Decay

With 'compound growth and decay', the amount that is added on or taken away at each stage is a percentage of the new amount — not the original amount.

Q1 Claire wants to sell her washing machine. She estimates that its value will have depreciated by 16% per year since she bought it. If Claire bought the washing machine two years ago for £220, how much should she sell it for (to the nearest £1)?

....

Q2 The population of squirrels in a forest has increased by 5% per year over the last four years. There were 168 squirrels in the forest four years ago. How many are there today?

....

Q3 Property prices in Angletown have depreciated by 10% per year over the last 3 years.

a) A house was worth £200 000 3 years ago. How much is it worth today?

....

b) Another house is worth £185 000 today. If house prices continue to depreciate by 10% per year, how many years will it take for the price of the house to fall below £150 000?

....

Q4 Ron buys a set of rare ornaments for £97. Their value increases by 4% in the first year after he buys them, 2% in the second year, and 5% in the third year. How much is the set worth after three years?

....

Q5 Chris has just bought a new laptop. He uses the formula $K = 550 \times 0.85^n$ to estimate what the value (£K) of his laptop will be in n years' time.

a) How much did Chris pay for the laptop? ....

b) How much will Chris's laptop be worth in 5 years? ....

> You'll be better off using the formula for the next two questions.
> Remember, it's $N = N_0 \times (\text{percentage change multiplier})^n$.
> N is the final amount, N_0 is the initial amount, and n is the number of time units (e.g. years).

Q6 A company predicts that the profit they make will increase at a rate of 1.5% per year. Their profit this year was £380 000.

If their prediction is correct, what will their profit be in 9 years' time? Give your answer to the nearest £1000. ....

Q7 Shaun invests £2000 into an account that pays 2.6% compound interest per annum. How much <u>interest</u> will he receive if he leaves his money in the account for 10 years?

....

Unit Conversions

You'll need to know the metric conversions for the exam, so get learning them. You don't need to memorise the conversions involving imperial units though — you'll be given the ones you need in the exam.

METRIC		IMPERIAL	METRIC-IMPERIAL
1 cm = 10 mm	1 tonne = 1000 kg	1 Yard = 3 feet	1 kg ≈ 2.2 pounds
1 m = 100 cm	1 litre = 1000 ml	1 Foot = 12 Inches	1 foot ≈ 30 cm
1 km = 1000 m	1 litre = 1000 cm³	1 Gallon = 8 Pints	1 gallon ≈ 4.5 litres
1 kg = 1000 g	1 cm³ = 1 ml	1 Stone = 14 Pounds	1 mile ≈ 1.6 km
		1 Pound = 16 Ounces	

Q1 Convert each of the following from kilograms to pounds.

10 kg ≈ .... lbs 16 kg ≈ .... lbs 75 kg ≈ .... lbs

Change each of these capacities from gallons to litres.

5 galls ≈ .... l 14 galls ≈ .... l 40 galls ≈ l

Q2 Complete the following conversions:

20 mm = ... cm .... mm = 6 cm 3470 m = ... km

.... m = 2 km 3 km = ... cm .... mm = 3.4 m

8550 g = ... kg 1.2 l = ... ml 4400 ml = ... l

Q3 A recipe for strawberry jam requires 4 lb of sugar. How many 1 kg bags of sugar does Sarah need to buy so that she can make the jam?

....

Q4 My garden shed is 235 cm tall. Approximately how many inches is this?

....

Q5 Complete each of the following imperial conversions.

8 stone = ... lbs 2.5 lbs = ... ounces 119 lbs = ... stone

40 pints = ... gallons 20 yards = ... feet 38.4 inches = ... feet

Unit Conversions

Q6 Justin is shopping online.
He looks up the following exchange rates:

1.60 US Dollars ($) to £1 Sterling.
175 Japanese Yen (¥) to £1 Sterling.
2 Australian Dollars (AUD) to £1 Sterling.

Use these exchange rates to calculate to the nearest penny the cost in Sterling of each of Justin's purchases:

a) A book costing $7.50

....

b) An MP3 player costing ¥7660

....

c) An electric guitar costing 683 AUD

....

d) Justin has two quotes for the cost of shipping his guitar from Australia to the UK:
155 AUD from an Australian courier and £76.45 from a British courier.
Which company is cheaper?

....

Q7 Tom walked 17 km in one day, while Ceara walked 10 miles. Who walked further?

....

It doesn't matter which distance you convert — but here it's easier to convert Ceara's miles to km.

Q8 Sophie is driving at 65 mph. The speed limit is 100 km/h.
How many km/h over the speed limit is Sophie driving?

....

Q9 David is throwing a party for himself and 15 of his friends.
He decides that it would be nice to make a bowl of fruit punch and carefully follows a recipe to make 1 gallon.

a) Will David fit all the punch in his 5 litre bowl?

....

Read the questions carefully — it's ml for part b), then pints for part c).

b) Is there enough punch for everyone at the party to have one 300 ml glass?

....

c) In the end, 12 people split the punch equally between them.
How many pints did they each drink?

....

Area and Volume Conversions

You've got to be really careful when you're converting areas and volumes — remember to divide/multiply by the conversion factor <u>twice</u> for <u>area</u> conversions and <u>three times</u> for <u>volume</u> conversions.

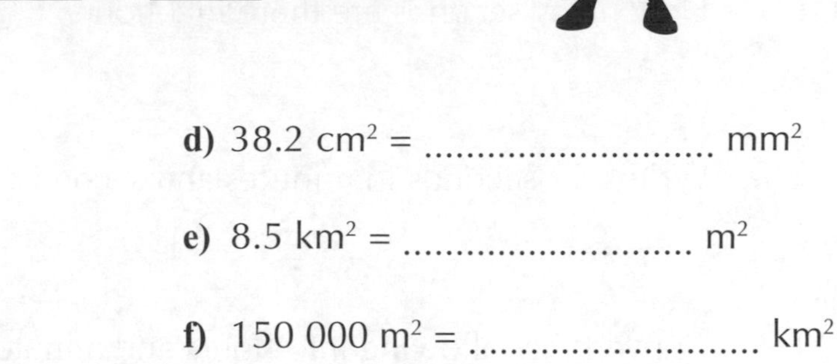

Q1 Convert these area measurements:

a) 900 mm^2 = ... cm^2

b) 4 m^2 = ... cm^2

c) 500 cm^2 = ... m^2

d) 38.2 cm^2 = ... mm^2

e) 8.5 km^2 = ... m^2

f) $150\,000 \text{ m}^2$ = ... km^2

Q2 Convert these volume measurements:

a) 4 cm^3 = ... mm^3

b) 3 m^3 = ... cm^3

c) $32\,500 \text{ mm}^3$ = ... cm^3

d) $55\,000 \text{ cm}^3$ = ... m^3

e) 25.1 cm^3 = ... mm^3

f) 8.3 m^3 = ... cm^3

Q3 A swimming pool has a volume of 172 m^3.
Convert this volume to cm^3.

....

Q4 Chris wants to varnish the surface of his table, which has an area of 1200 cm^2.
He buys a tin of varnish that says it will cover an area of 2 m^2.

Find the area of the table surface in m^2.
Does Chris have enough varnish to cover the surface of his table?

....

Q5 Mandy is making herself and two of her friends a cup of tea.
For each cup of tea she needs $300\,000 \text{ mm}^3$ of water.

Don't forget to multiply the amount she needs for each cup by three...

She fills her kettle with 850 cm^3 of water. Is this enough water to make the tea?

....

Time Intervals

I know time questions might sound easy, but don't get tripped up by decimals — 1.25 hours means 1 hour 15 minutes, not 1 hour 25 minutes.

Q1 How many seconds are there in 4 hours? ....

Q2 Write 212 seconds in minutes and seconds. ....

Q3 Convert the following into hours and minutes:

a) 3.75 hours .... **b)** 0.2 hours .... **c)** 5.8 hours ....

Q4 It takes Jeremy 4 hours to drive from his house to his parents' house.
It takes him a third of this time to drive from his house to his brother's house.

How long does it take him to drive to his brother's house?
Give your answer in hours and minutes. ....

Q5 Steve sets off on a bike ride at 10.30 am. He stops for lunch at 12.15 pm and sets off again at 1 pm. He has a 20 minute rest stop in the afternoon and gets home at 4.50 pm.

How long did he cycle for altogether? ....

Q6 Fran is going to cook a 5.5 kg ham. The ham needs to be cooked for 40 minutes per kg, and then left to "rest" for 25 minutes before serving. If Fran wants to serve the ham at 4 pm, what time should she start cooking?

....

Q7 This timetable refers to three trains that travel from Asham to Derton.

a) Which train is quickest from Asham to Derton? ....

b) Which train is quickest from Cottingham to Derton?

....

c) I live in Bordhouse. It takes me 8 minutes to walk to the train station. What time must I leave the house by to arrive in Derton before 2.30 pm?

Asham – Derton	Train 1	Train 2	Train 3
Asham	0832	1135	1336
Bordhouse	0914	1216	1414
Cottingham	1002	1259	1456
Derton	1101	1404	1602

....

Speed

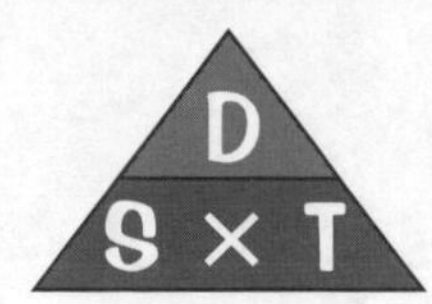

To work out a speed, distance or time, you can use the good old formula triangle. Just make sure you can remember the order of the letters — S D T.

Q1 A motorbike travels for 3 hours at an average speed of 55 mph. How far has it travelled?

....

Q2 Complete this table.

Distance Travelled	Time taken	Average Speed
210 km	3 hrs	
135 miles		30 mph
	2 hrs 30 mins	42 km/h
9 miles	45 mins	
640 km		800 km/h
	1 hr 10 mins	60 mph

Q3 A snail travels at a speed of 2.8×10^{-4} m/s.
How long would it take the snail to travel 3.5×10^{-2} m?
Give your answer in minutes and seconds.

....

Q4 An athlete can run 100 m in 11 seconds.
Calculate the athlete's speed in:

a) m/s ....

b) km/h ....

Q5 Simon is driving on the motorway at an average speed of 67 mph.
He sees a sign telling him that he is 36 miles away from the next service station.

a) To the nearest minute, how long will it take Simon to reach the service station?

....

b) Later, Simon drove through some roadworks where the speed limit was 50 mph.
Two cameras recorded the time taken to travel 1200 m through the roadworks as 56 seconds. Was Simon speeding through the roadworks?

....

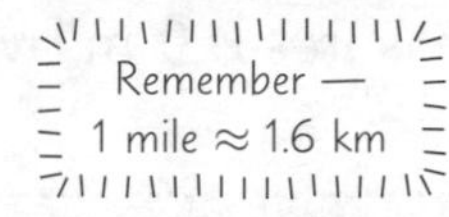

Density and Pressure

M
D × V

You'll need another couple of formula triangles here. Get them learnt and keep an eye on the units and everything'll be fine and dandy.

F
P × A

Q1 Find the density of each of these pieces of wood. Give your answer in g/cm^3:

a) Mass 3 g, volume 4 cm^3 ....

b) Mass 20 g, volume 25 cm^3 ....

Q2 Calculate the mass of each of these objects:

a) a small marble statue of density 2.6 g/cm^3 and volume 24 cm^3 ....

b) a plastic cube of volume 64 cm^3 and density 1.5 g/cm^3 ....

Q3 Work out the volume of each of these items:

a) a bag of sugar of mass 1000 g and density 1.6 g/cm^3 ....

b) a packet of margarine with density 2.8 g/cm^3 and mass 250 g ....

Q4 A gold bar measures 12 cm by 4 cm by 4 cm and has density 19.5 g/cm^3. Calculate the mass of the gold bar.

....

Q5 A solid rubber ball has mass 0.24 kg and volume 200 cm^3. What is its density in g/cm^3?

....

Q6 A block of ice has mass 18.6 kg and volume 0.02 m^3. What is its density in g/cm^3?

....

Q7 A mattress weighing 450 N sits on the floor, which is horizontal. The side of the mattress resting on the floor has an area of 3 m^2. Work out the pressure exerted by the mattress on the floor.

Remember, weight is a force, so it's measured in newtons.

....

Q8 When a chest of drawers of weight 680 N rests on horizontal ground, it exerts a pressure of 850 N/m^2. Work out the area of the base of the chest of drawers.

....

Q9 A large speaker sitting on horizontal ground exerts 640 N/m^2 of pressure on the ground. The side of the speaker that rests on the ground is rectangular and measures 0.9 m by 0.8 m. What is the weight of the speaker?

....

Properties of 2D Shapes

To work out if you've got a line of symmetry, just imagine you're folding the shape in half. If the sides will fold exactly together, then hey presto, it's symmetrical about the fold.

Q1 These shapes have more than one line of symmetry.
Draw the lines of symmetry using dotted lines.

a)

b)

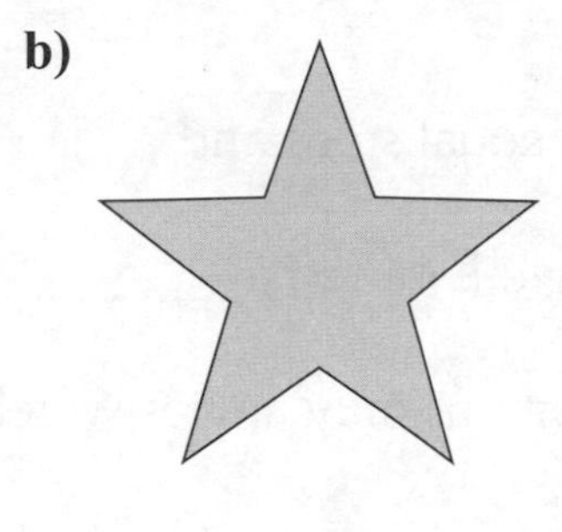

c)

Q2 Complete the following diagrams so that they have rotational symmetry about centre C of the order stated:

a) order 2 **b)** order 4 **c)** order 3

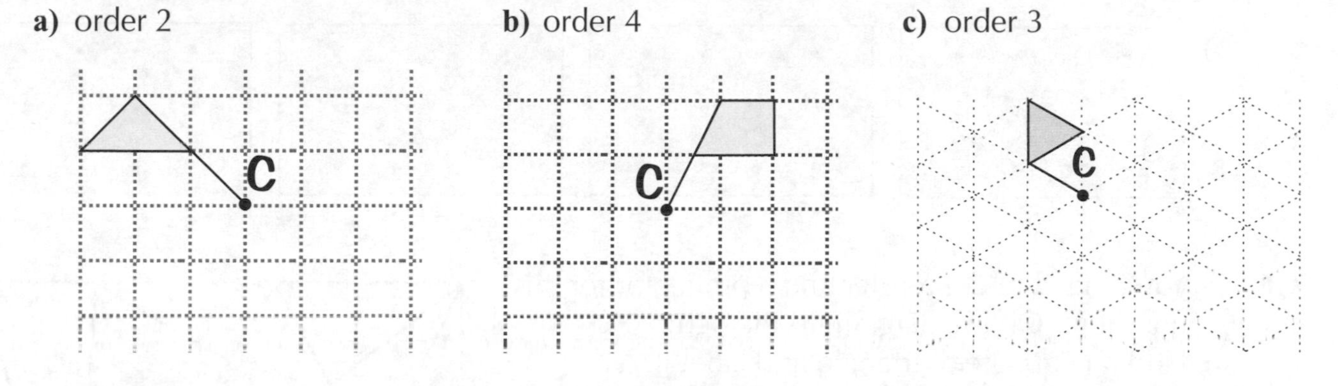

Q3 Name each of these regular polygons, and then write down the number of lines of symmetry and the order of rotational symmetry in the spaces below.

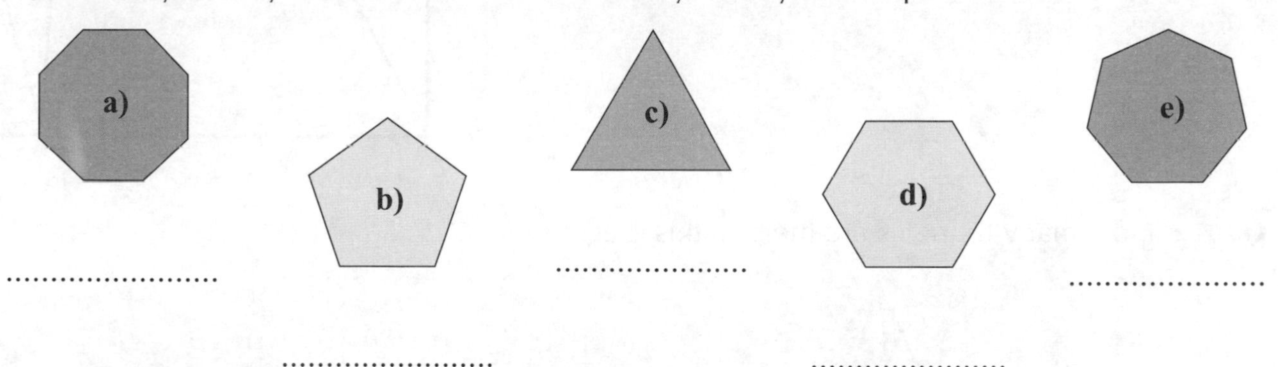

a) Lines of symmetry = ... Order of rotational symmetry = ...

b) Lines of symmetry = ... Order of rotational symmetry = ...

c) Lines of symmetry = ... Order of rotational symmetry = ...

d) Lines of symmetry = ... Order of rotational symmetry = ...

e) Lines of symmetry = ... Order of rotational symmetry = ...

Properties of 2D Shapes

Q4 Fill in the gaps in these sentences.

a) An isosceles triangle has .... equal sides and .... equal angles.

b) A triangle with all its sides equal and all its angles equal is called an

.... triangle.

c) A scalene triangle has .... equal sides and .... equal angles.

d) A triangle with one right angle is called a .... triangle.

e) An acute-angled triangle contains only angles which are .... than ....°.

f) A triangle with one angle greater than 90° is called an .... triangle.

Q5 By joining dots, draw four different isosceles triangles — one in each box.

Q6 You'll need to use a ruler and a protractor for this question. On the diagram to the right:

a) Find an equilateral triangle and label it 'A'.

b) Label two different right-angled triangles 'B'.

c) Label two different scalene triangles 'C'.

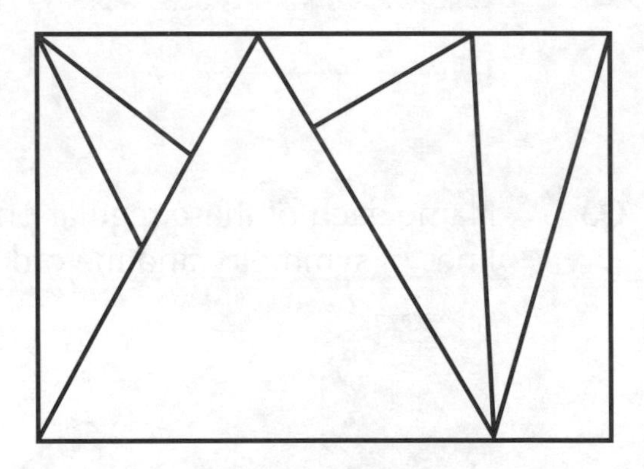

Q7 How many triangles are there in this diagram?

....

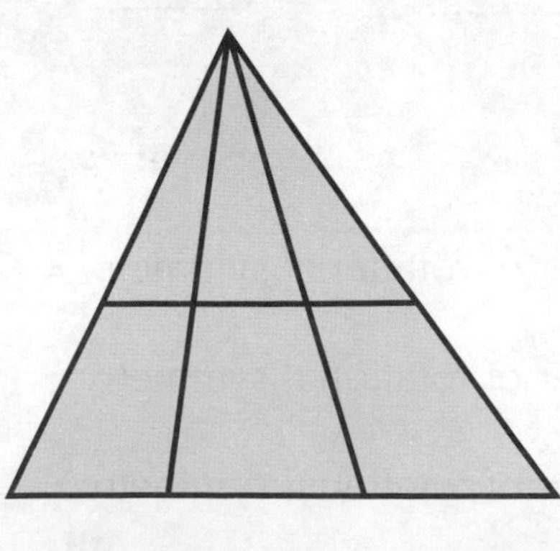

Try counting the triangles a few times — there are more than you might think...

What sort of triangles are they?

....

There are only a few types of triangle, so make sure you know them all — think of all those nice fat juicy marks...

Properties of 2D Shapes

Here are a few easy marks for you — all you've got to do is remember the shapes and a few facts about them... it's a waste of marks not to bother.

Q8 Fill in the blanks in the table.

NAME	DRAWING	DESCRIPTION
Square		All sides of equal length. Opposite sides parallel. Four right angles.
....	....	Two pairs of equal-length, parallel sides. Four right angles.
....		Opposite sides are .... and equal. Opposite angles are equal.
Trapezium		Only .... sides are parallel.
Rhombus		A parallelogram but with all sides ....
Kite	....	Two pairs of adjacent equal sides.

Q9 Stephen is measuring the angles inside a parallelogram for his maths homework. To save time, he measures just one and works out what the other angles must be from this. If the angle he measures is 52°, what are the other three?

.........., ..., ...

Congruent Shapes

Remember the four conditions for triangles to be congruent — SSS, AAS, SAS and RHS (you only need to show that one of these is true, not all four).

Q1 For the following set of shapes, underline the one which is <u>not</u> congruent to the others.

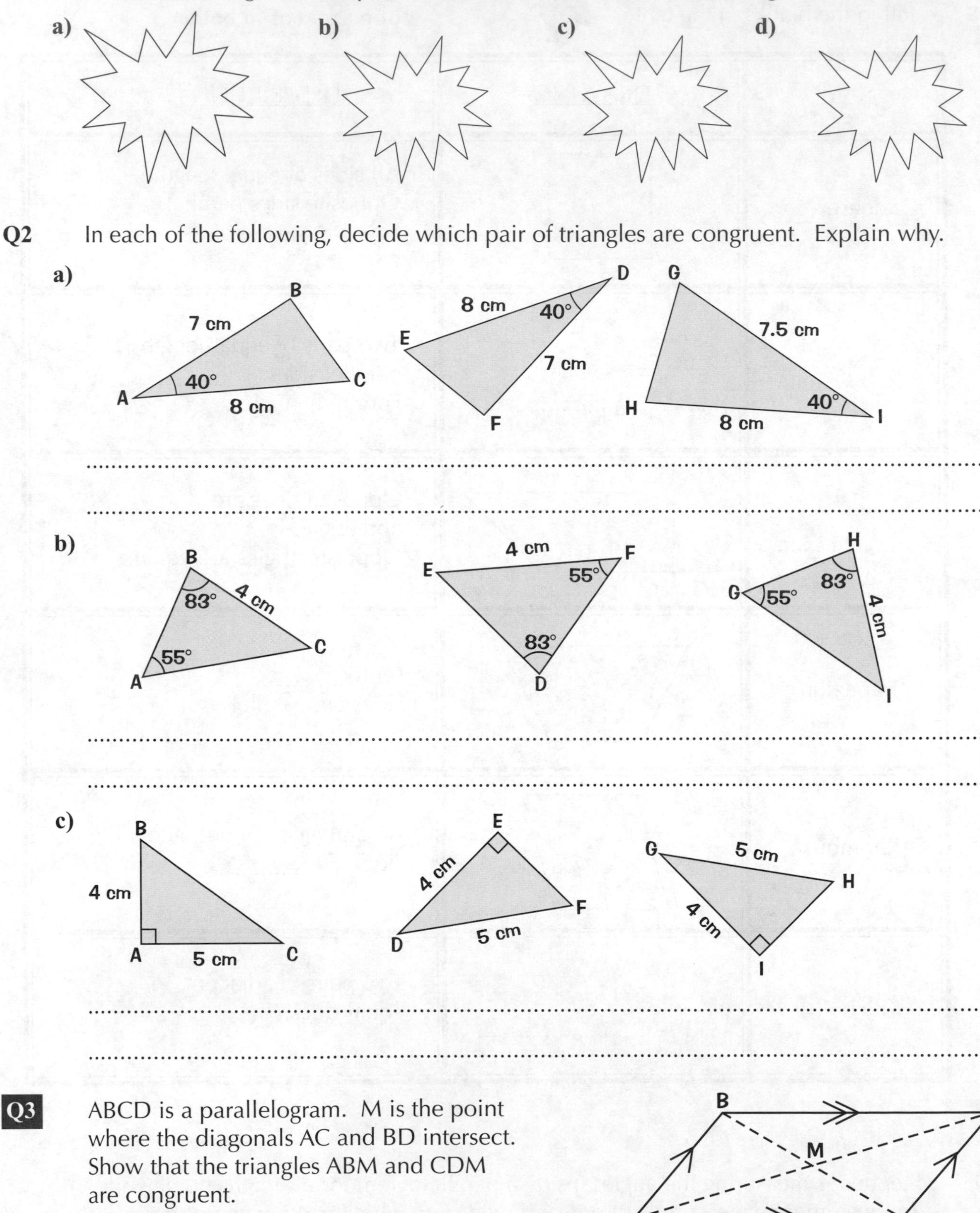

Q2 In each of the following, decide which pair of triangles are congruent. Explain why.

a) ....

....

b) ....

....

c) ....

....

Q3 ABCD is a parallelogram. M is the point where the diagonals AC and BD intersect. Show that the triangles ABM and CDM are congruent.

....

....

....

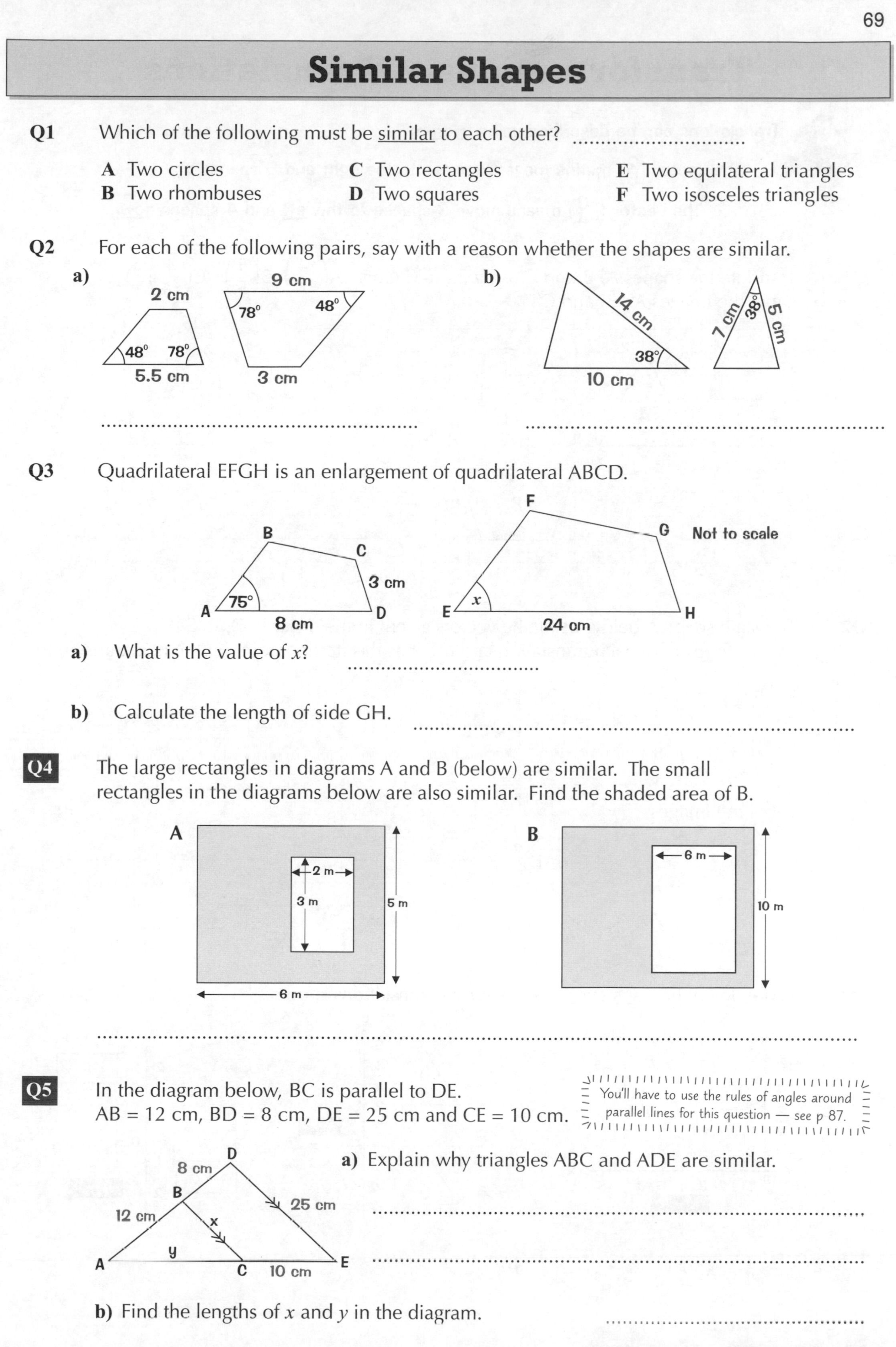

Similar Shapes

Q1 Which of the following must be similar to each other? ....

A Two circles
B Two rhombuses
C Two rectangles
D Two squares
E Two equilateral triangles
F Two isosceles triangles

Q2 For each of the following pairs, say with a reason whether the shapes are similar.

a)

....

b)

....

Q3 Quadrilateral EFGH is an enlargement of quadrilateral ABCD.

a) What is the value of x? ....

b) Calculate the length of side GH. ....

Q4 The large rectangles in diagrams A and B (below) are similar. The small rectangles in the diagrams below are also similar. Find the shaded area of B.

....

Q5 In the diagram below, BC is parallel to DE.
AB = 12 cm, BD = 8 cm, DE = 25 cm and CE = 10 cm.

You'll have to use the rules of angles around parallel lines for this question — see p 87.

a) Explain why triangles ABC and ADE are similar.

....

....

b) Find the lengths of x and y in the diagram. ....

Transformations — Translations

Translations can be described using vectors.

The vector $\begin{pmatrix} 2 \\ 5 \end{pmatrix}$ means move 2 spaces to the right and 5 spaces up.

The vector $\begin{pmatrix} -3 \\ -4 \end{pmatrix}$ means move 3 spaces to the left and 4 spaces down.

Q1 Translate the shapes A, B and C using these vectors: A$\begin{pmatrix} -4 \\ -3 \end{pmatrix}$ B$\begin{pmatrix} 5 \\ 5 \end{pmatrix}$ C$\begin{pmatrix} 4 \\ -4 \end{pmatrix}$
Label the images A′, B′ and C′.

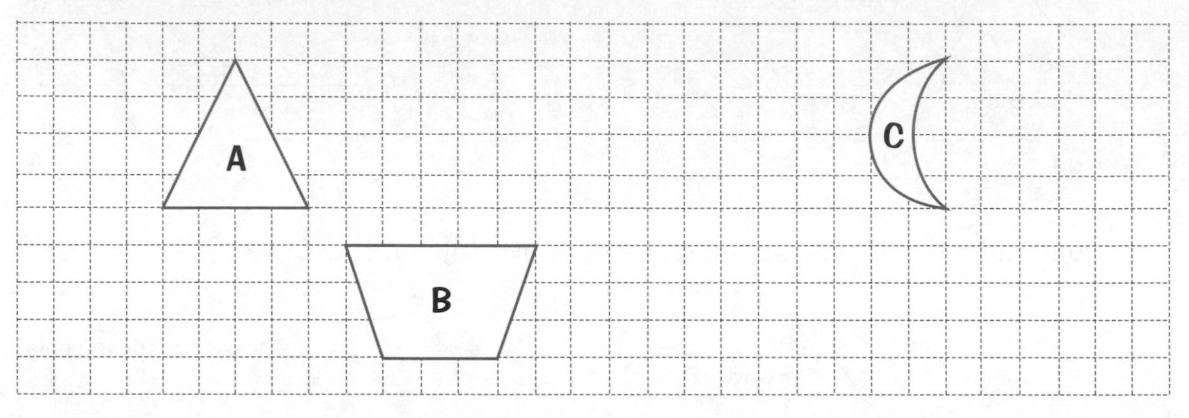

Q2 Translate shape A below using the vectors given, in the order P, Q, R, S.
Draw the result of each translation and use it as the starting point for the next translation.

P$\begin{pmatrix} 3 \\ 4 \end{pmatrix}$ Q$\begin{pmatrix} 9 \\ 2 \end{pmatrix}$

R$\begin{pmatrix} 3 \\ -4 \end{pmatrix}$ S$\begin{pmatrix} -8 \\ -4 \end{pmatrix}$

Label the images A′, A″ , A‴, A⁗.

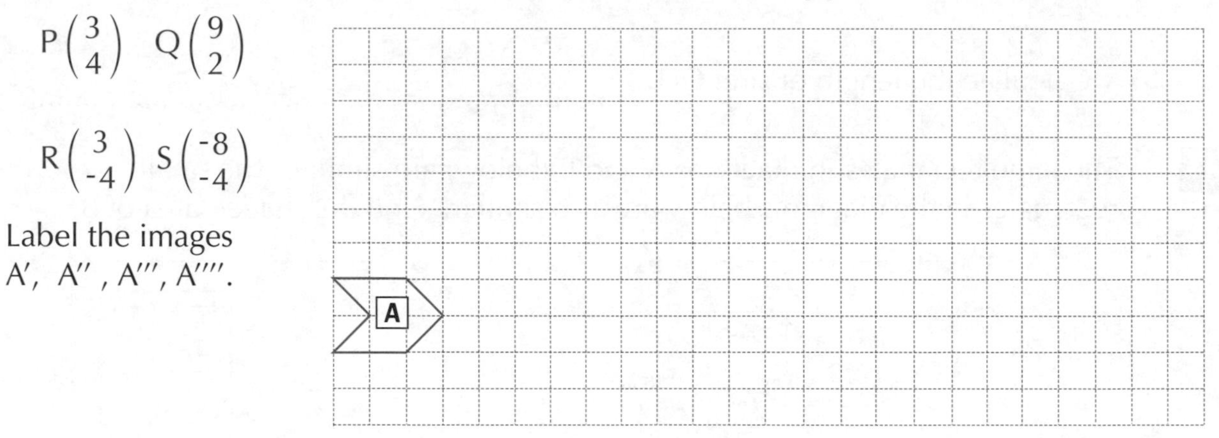

Q3 Write down the translation vectors for the translations shown.

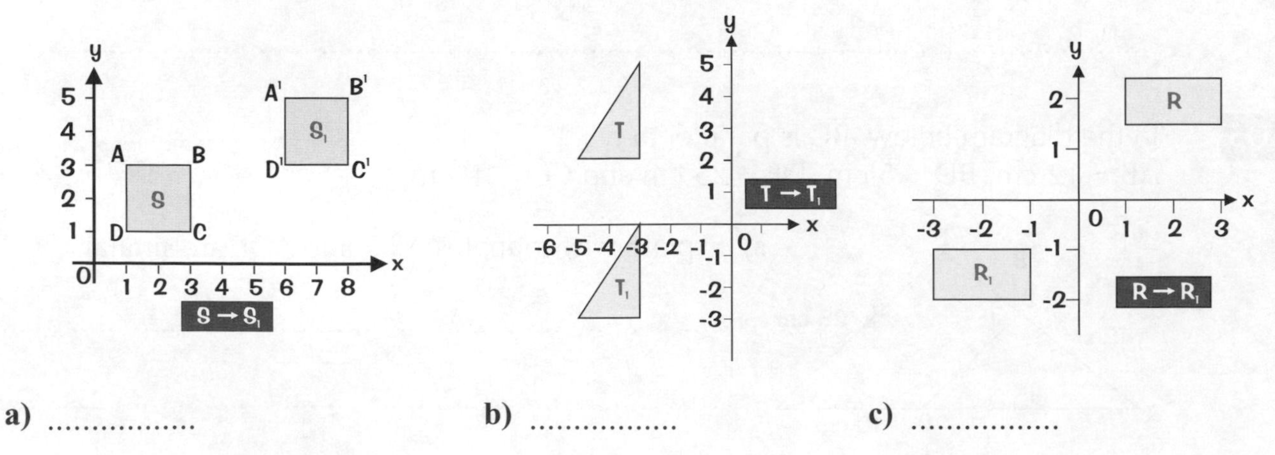

a) .... **b)** .... **c)** ....

Transformations — Rotations

A 180° rotation clockwise is the same as a 180° rotation anticlockwise — and a 90° rotation clockwise is the same as a 270° rotation anticlockwise. Great fun, innit...

Q1 The scalene triangle PQR is shown below. It is rotated 90° anticlockwise about the origin to give the new triangle P′Q′R′.

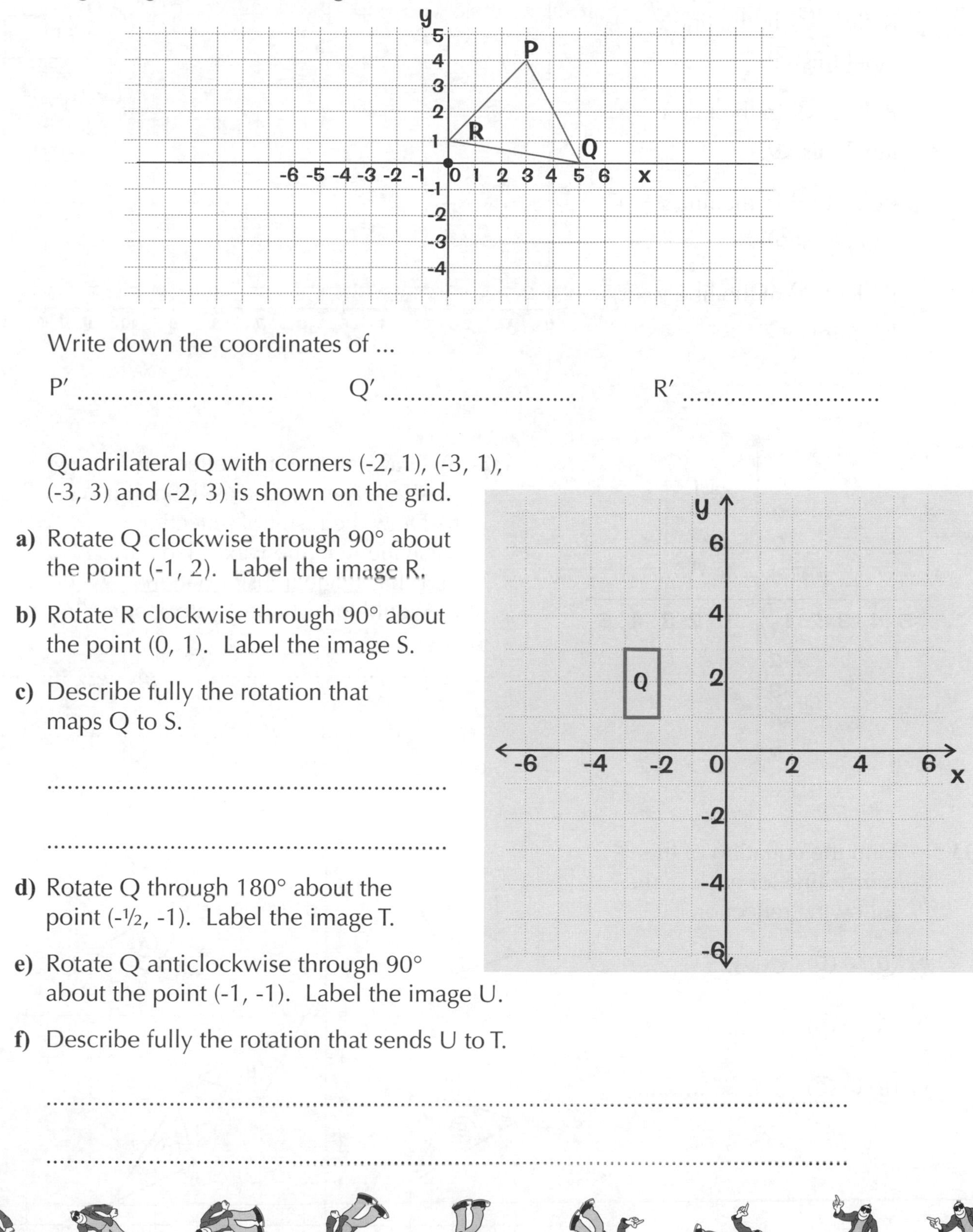

Write down the coordinates of ...

P′ .... Q′ .... R′ ....

Q2 Quadrilateral Q with corners (-2, 1), (-3, 1), (-3, 3) and (-2, 3) is shown on the grid.

a) Rotate Q clockwise through 90° about the point (-1, 2). Label the image R.

b) Rotate R clockwise through 90° about the point (0, 1). Label the image S.

c) Describe fully the rotation that maps Q to S.

....

....

d) Rotate Q through 180° about the point (-½, -1). Label the image T.

e) Rotate Q anticlockwise through 90° about the point (-1, -1). Label the image U.

f) Describe fully the rotation that sends U to T.

....

....

Transformations — Reflections

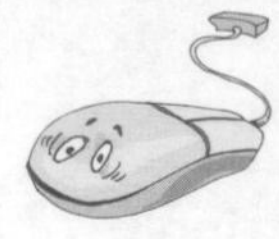

Nothing fancy here. Reflections are just mirror drawing really. And we've all done that before...

Q1 Reflect ① in the line $y = 5$, label this ②.

Reflect ② in the line $x = 9$, label this ③.

Reflect ③ in the line $y = x$, label this ④.

Reflect ④ in the line $x = 4$, label this ⑤.

Reflect ⑤ in the line $y = x$, label this ⑥.

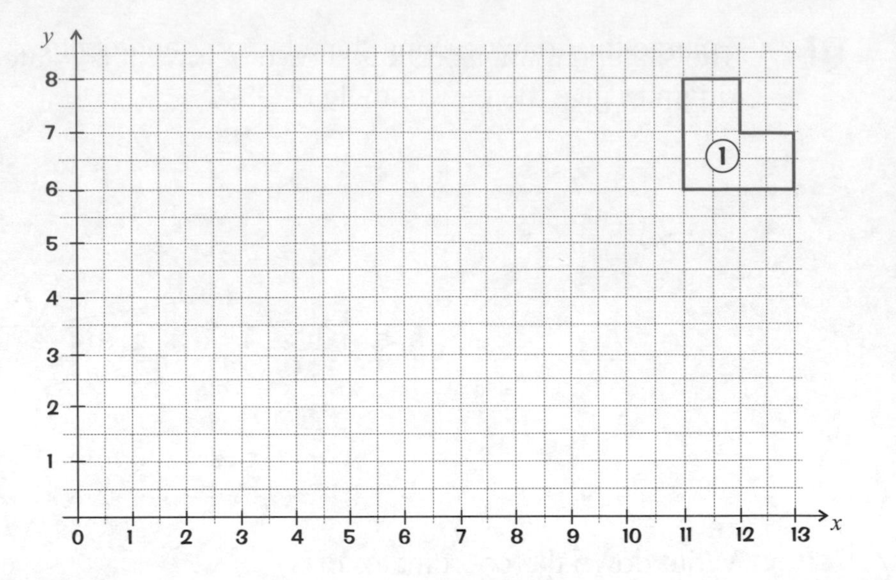

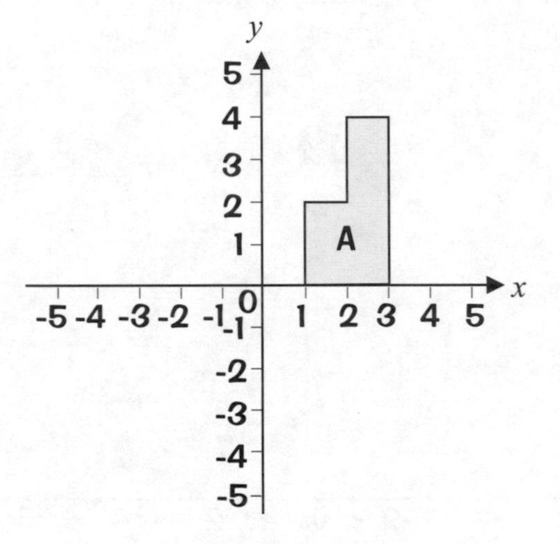

Q2 **a)** Draw the result of reflecting shape A in the x-axis, label this A′.

b) Draw the result of reflecting shape A′ in the y-axis, label this A″.

c) What single transformation would turn shape A into shape A″?

....

....

Q3 Find the equation of the mirror line for each of the following reflections:

a) Ⓓ to Ⓔ ....

b) Ⓐ to Ⓓ ....

c) Ⓑ to Ⓒ ....

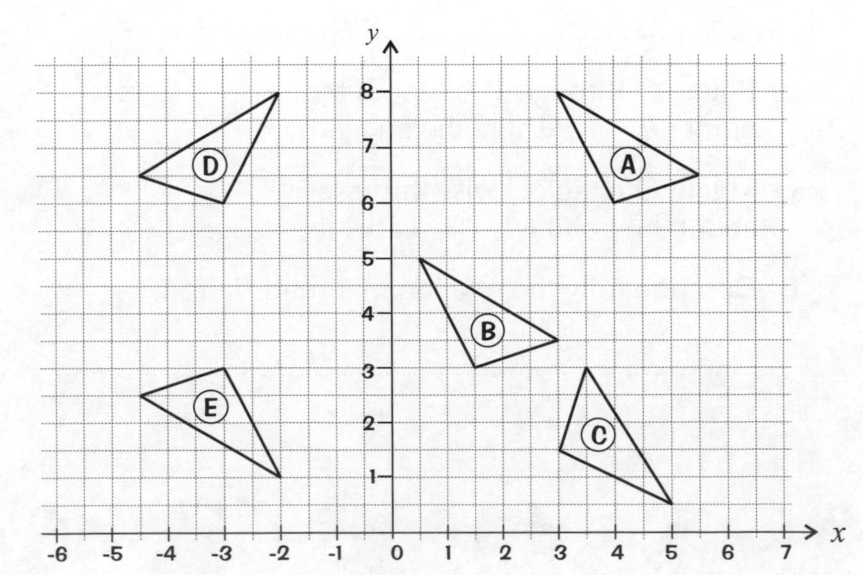

Transformations — Enlargements

The scale factor is a fancy way of saying HOW MUCH BIGGER the enlargement is than the original — e.g. a scale factor of 3 means it's <u>3 times as big</u>.

Q1 Enlarge this triangle using scale factor 2 and centre of enlargement C.

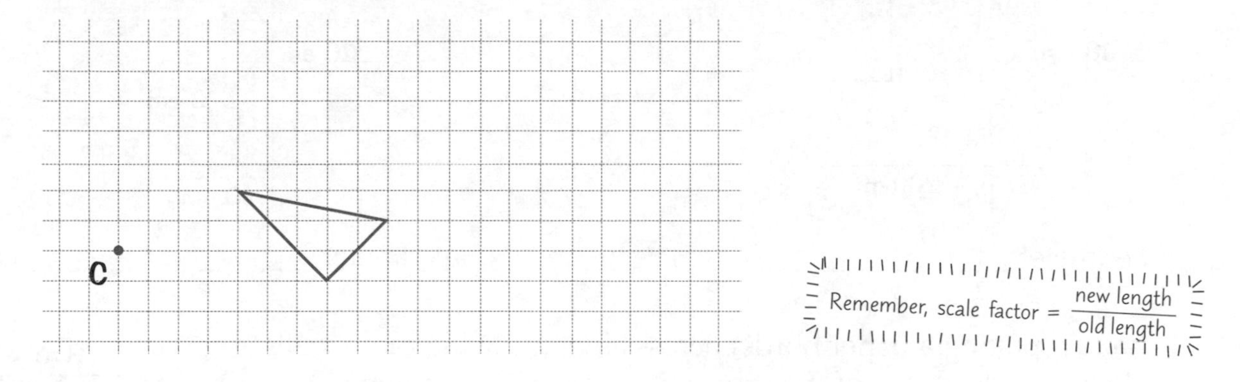

Q2 a) What is the scale factor of each of these enlargements?

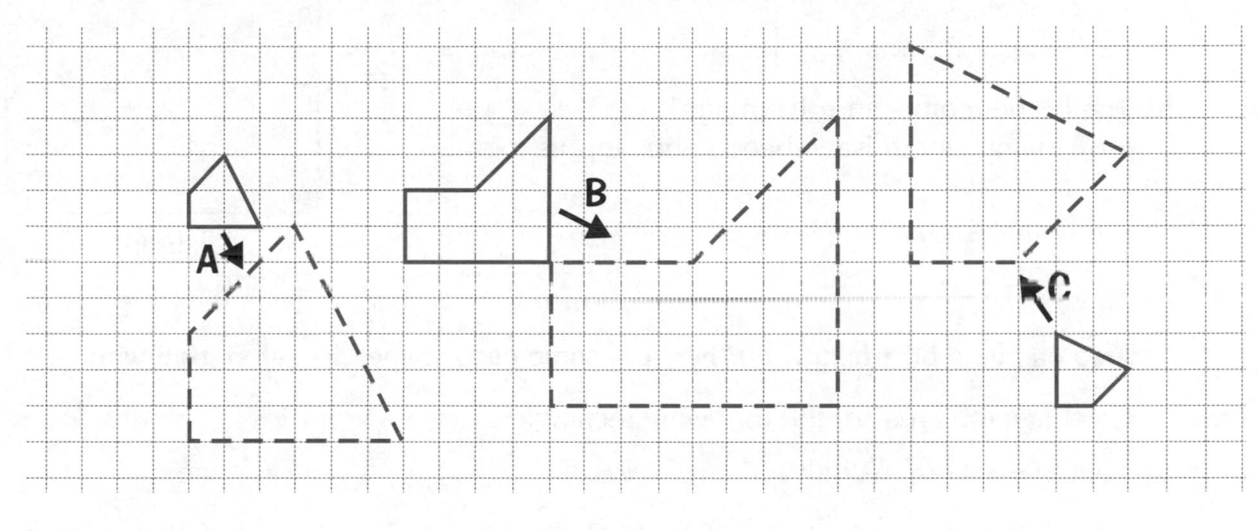

A: Scale factor is .... **B:** Scale factor is .... **C:** Scale factor is ....

b) Mark the centre of enlargement for each of the shapes above.

Q3 Enlarge this shape using scale factor $\frac{1}{3}$ and centre of enlargement (-7, 1).

y: -3 -2 -1 0 1 2 3 4
x: -8 -7 -6 -5 -4 -3 -2 -1 0 1 2 3 4 5 6 7 8 9 10

Perimeter and Area

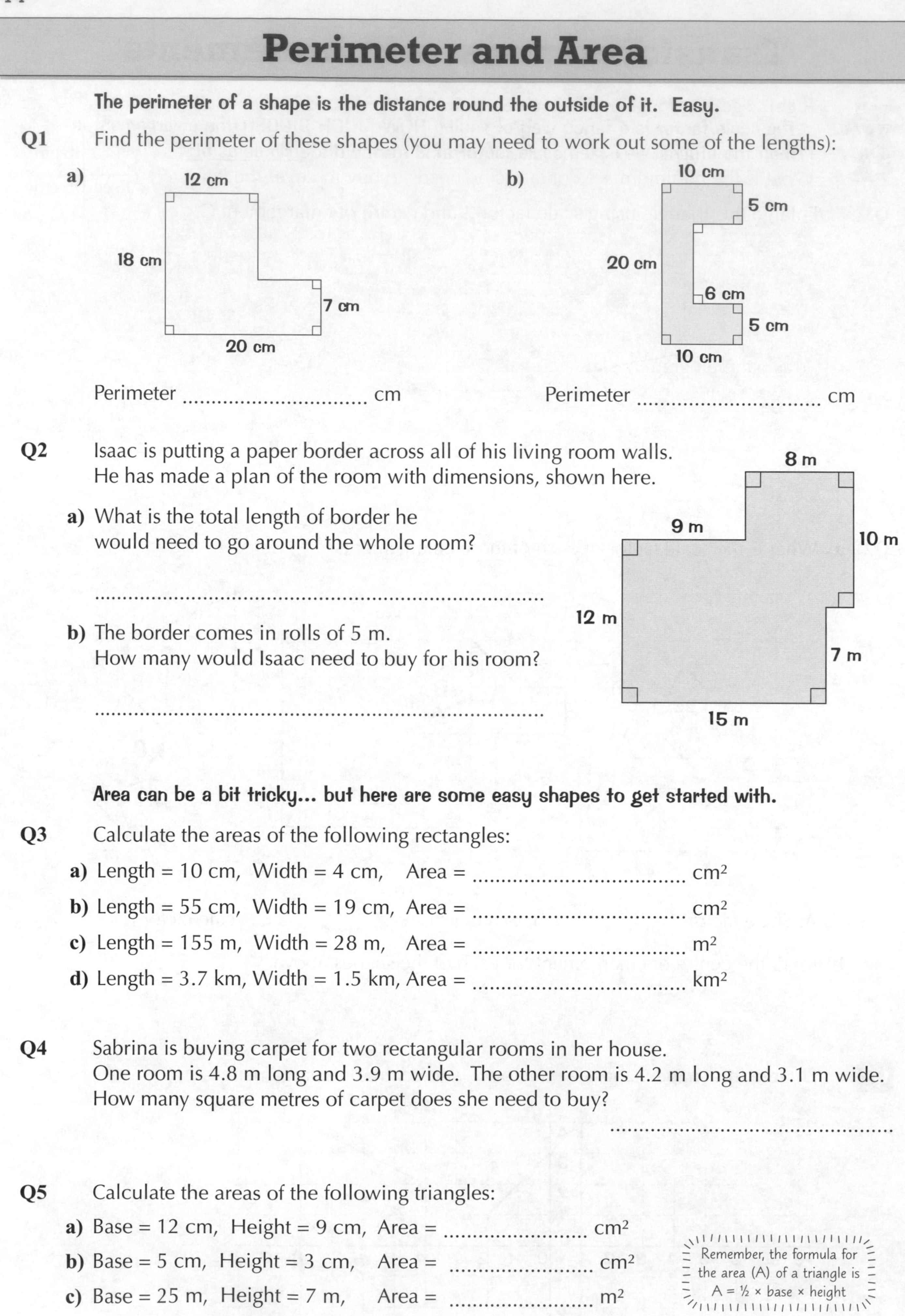

The perimeter of a shape is the distance round the outside of it. Easy.

Q1 Find the perimeter of these shapes (you may need to work out some of the lengths):

a)

Perimeter .... cm

b)

Perimeter .... cm

Q2 Isaac is putting a paper border across all of his living room walls. He has made a plan of the room with dimensions, shown here.

a) What is the total length of border he would need to go around the whole room?

....

b) The border comes in rolls of 5 m. How many would Isaac need to buy for his room?

....

Area can be a bit tricky... but here are some easy shapes to get started with.

Q3 Calculate the areas of the following rectangles:

a) Length = 10 cm, Width = 4 cm, Area = ... cm²

b) Length = 55 cm, Width = 19 cm, Area = ... cm²

c) Length = 155 m, Width = 28 m, Area = ... m²

d) Length = 3.7 km, Width = 1.5 km, Area = ... km²

Q4 Sabrina is buying carpet for two rectangular rooms in her house. One room is 4.8 m long and 3.9 m wide. The other room is 4.2 m long and 3.1 m wide. How many square metres of carpet does she need to buy?

....

Q5 Calculate the areas of the following triangles:

a) Base = 12 cm, Height = 9 cm, Area = ... cm²

b) Base = 5 cm, Height = 3 cm, Area = ... cm²

c) Base = 25 m, Height = 7 m, Area = ... m²

d) Base = 1.6 m, Height = 6.4 m, Area = ... m²

Remember, the formula for the area (A) of a triangle is A = ½ × base × height

Perimeter and Area

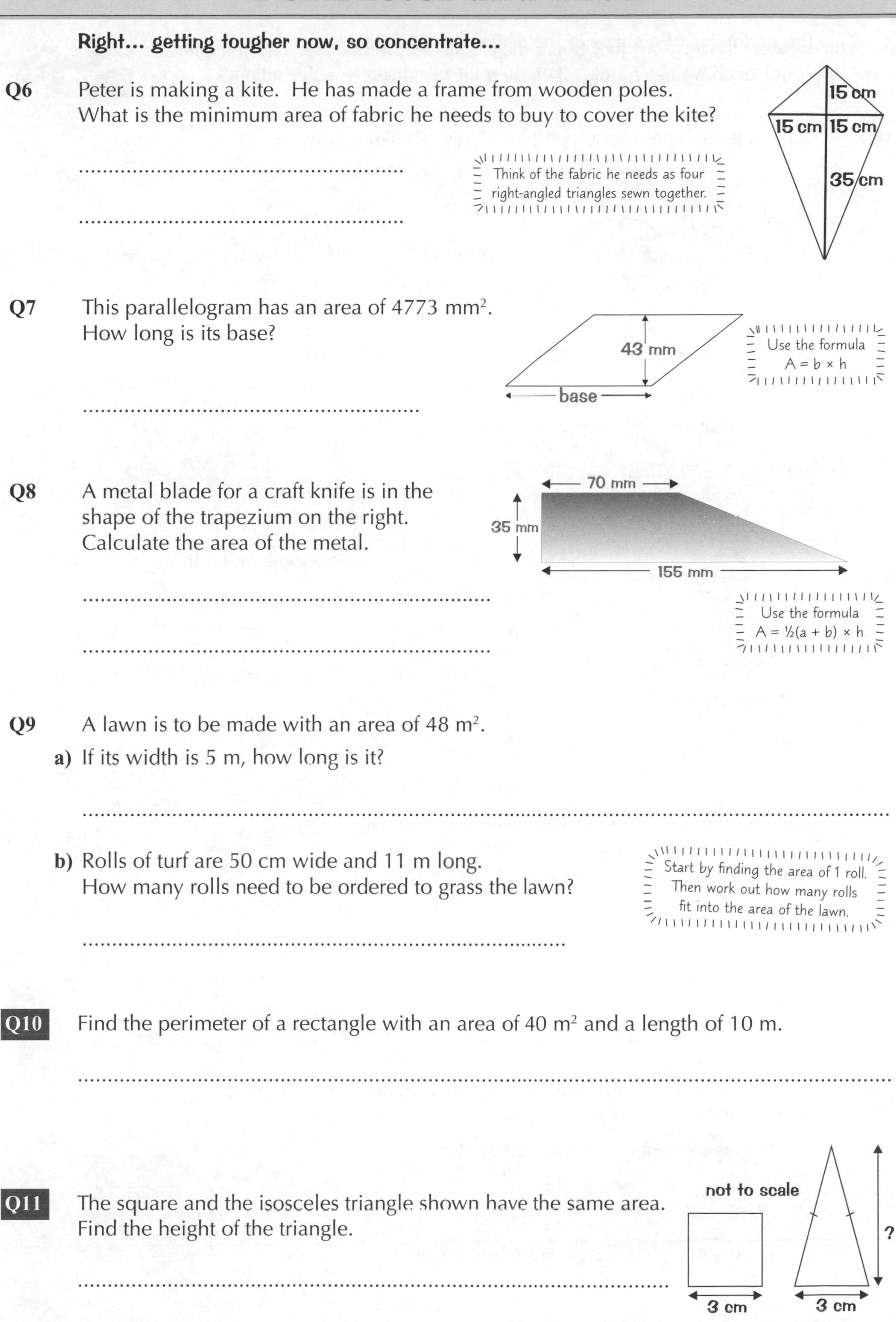

Right... getting tougher now, so concentrate...

Q6 Peter is making a kite. He has made a frame from wooden poles.
What is the minimum area of fabric he needs to buy to cover the kite?

....

....

Think of the fabric he needs as four right-angled triangles sewn together.

Q7 This parallelogram has an area of 4773 mm^2.
How long is its base?

....

Use the formula $A = b \times h$

Q8 A metal blade for a craft knife is in the shape of the trapezium on the right.
Calculate the area of the metal.

....

....

Use the formula $A = ½(a + b) \times h$

Q9 A lawn is to be made with an area of 48 m^2.

a) If its width is 5 m, how long is it?

....

b) Rolls of turf are 50 cm wide and 11 m long.
How many rolls need to be ordered to grass the lawn?

....

Start by finding the area of 1 roll. Then work out how many rolls fit into the area of the lawn.

Q10 Find the perimeter of a rectangle with an area of 40 m^2 and a length of 10 m.

....

Q11 The square and the isosceles triangle shown have the same area.
Find the height of the triangle.

....

Perimeter and Area

Bit trickier, these... but just break the shapes into rectangles and triangles. Work out each bit separately, then do a bit of adding or subtracting.

Q12 Calculate the areas of the complex shapes shown below.

a)

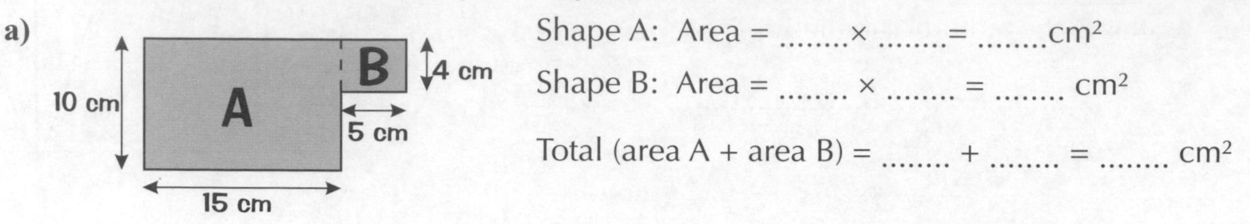

Shape A: Area = ... × .... = ...cm²

Shape B: Area = ... × .... = ... cm²

Total (area A + area B) = ... + ... = ... cm²

b) The shape on the right is a rectangle with a triangular piece removed.

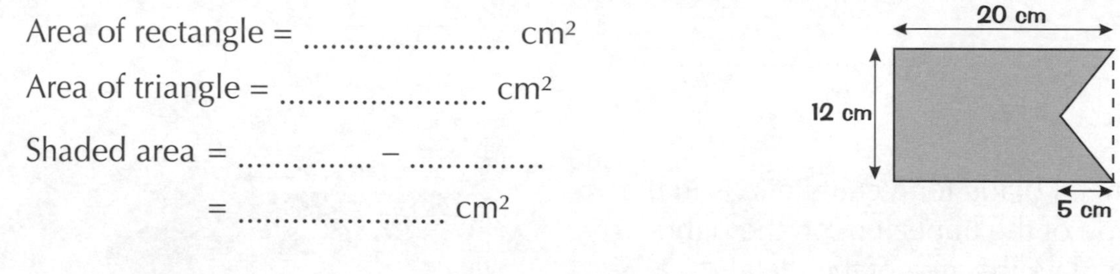

Area of rectangle = ... cm²

Area of triangle = ... cm²

Shaded area = ... − ....

= .... cm²

c)

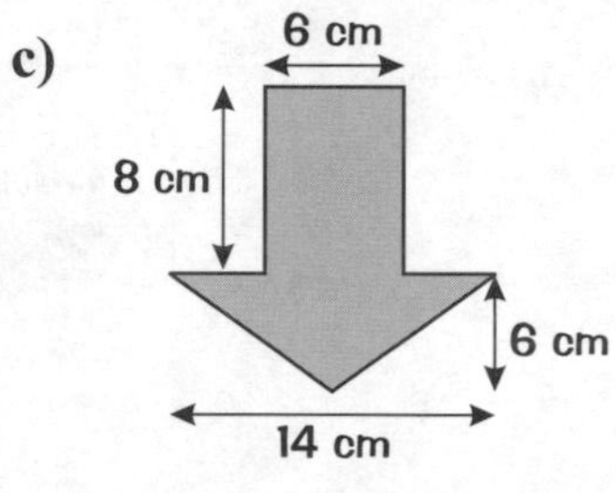

Divide this shape into a rectangle and a triangle.

Area of rectangle = ... cm²

Area of triangle = ... cm²

Area of whole shape = ... + ...

= .... cm²

Q13 Tariq is making a gravelled patio in the shape shown. He can buy gravel in bags that cover 1 m² each. How many bags of gravel does he need to buy?

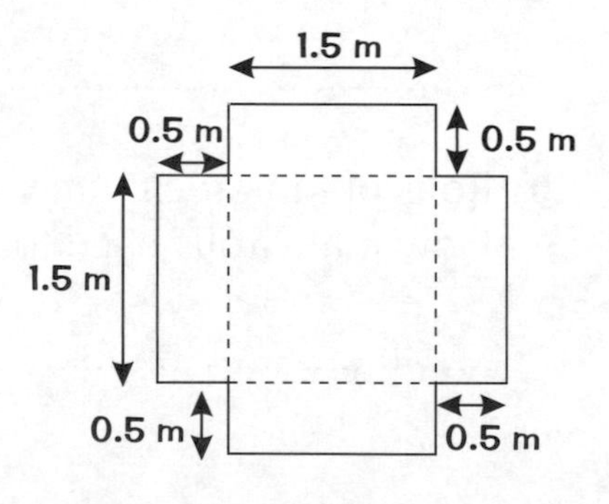

....

....

Q14 The diagram on the right is made up of one square inside another. All four white triangles are congruent.

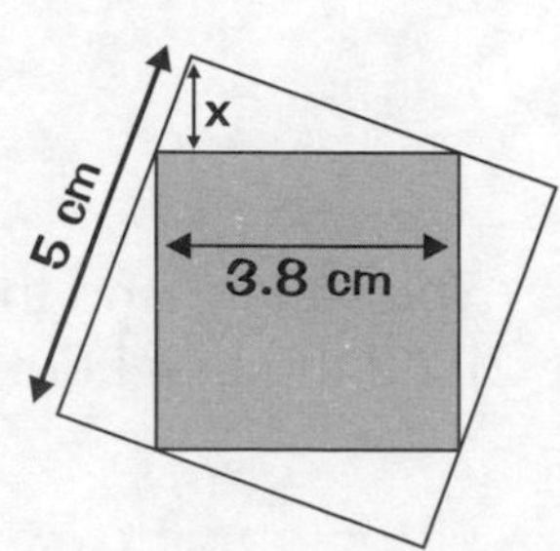

a) Find the area of one of the white triangles.

....

b) Find the distance labelled x, correct to 1 d.p.

....

Perimeter and Area — Circles

Don't worry about that π symbol — it just stands for the number 3.14159...
Your calculator should have a handy π button you can use.
If you're not allowed a calculator, just use 3.142.

Q1 A minor sector is labelled A on the diagram.
Name the features labelled:

B: ....

C: .... D: ....

Q2 Calculate the circumference and area of these circles.
Give your answers to 2 d.p.

a) 2 cm

Circumference = ...

Area = ...

b) 2.5 cm

Circumference = ...

Area = ...

Q3 What is the area of this semicircular rug?
Give your answer to the nearest whole number.

150 cm

....

Q4 A wheel has a diameter of 0.6 m. How far does it travel in one complete turn?
Give your answer to 2 d.p.

....

Q5 Patrick is using weed killer to treat his circular lawn, which has a diameter of 12 m.
The weed killer instructions say to use 20 ml for every square metre of lawn to be treated.

a) What is the area of Patrick's lawn?

....

b) How much weed killer does he need to use? Give your answer to the nearest ml.

....

Q6 A bicycle wheel has a radius of 32 cm. Calculate how far the bicycle will have travelled after the wheel has turned 100 times. Give your answer to the nearest metre.

....

....

You only need to work out the distance covered by one of the wheels, as the bike will travel the same distance.

Perimeter and Area — Circles

Oh go on then, I'll give you the formula for the length of an arc — it's arc length = $\frac{x}{360}$ × circumference of full circle. While I'm at it, I might as well tell you that the area of a sector = $\frac{x}{360}$ × area of full circle.

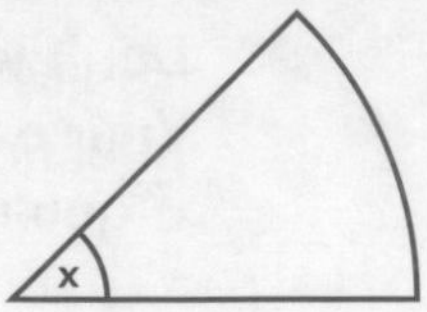

Q7 For each of the following, give your answers in terms of π.

a) Find the length of the arc shown on the right.

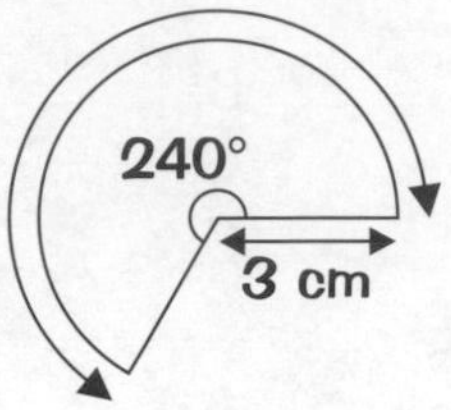

240°
3 cm

....

....

b) Find the area of the sector shown on the right.

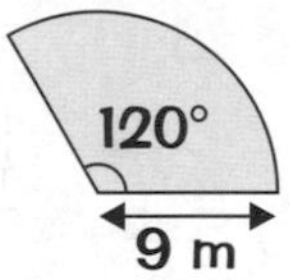

120°
9 m

....

....

Q8 Emily has a wedge-shaped piece of cheese, as shown below.
She traces around her cheese onto a piece of paper to create a sector.

a) i) How long is the curved part of her outline? Give your answer to 2 decimal places.

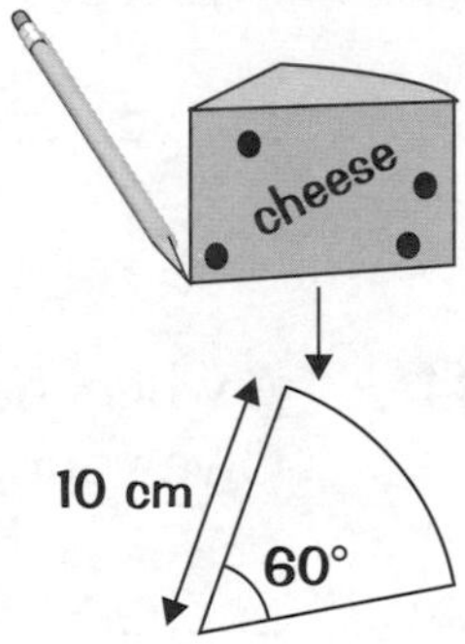

....

....

cheese
10 cm
60°

ii) What is the total perimeter of the base of her piece of cheese?

....

....

b) What is the area of the shape Emily has drawn?

....

....

Q9 The diagram on the right shows a tile with a design based on four quarter circles inside a square.
The quarter circles all have the same radius.

Find the area of the white part of the tile, to 2 d.p.

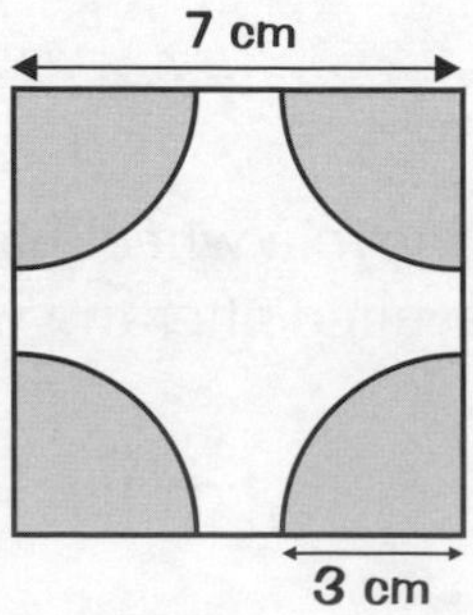

7 cm
3 cm

....

3D Shapes — Surface Area

You're in for some lovely pages about 3D shapes. Try answering these questions while wearing 3D specs... they'll probably seem easier.

Q1 What are the names of these shapes?

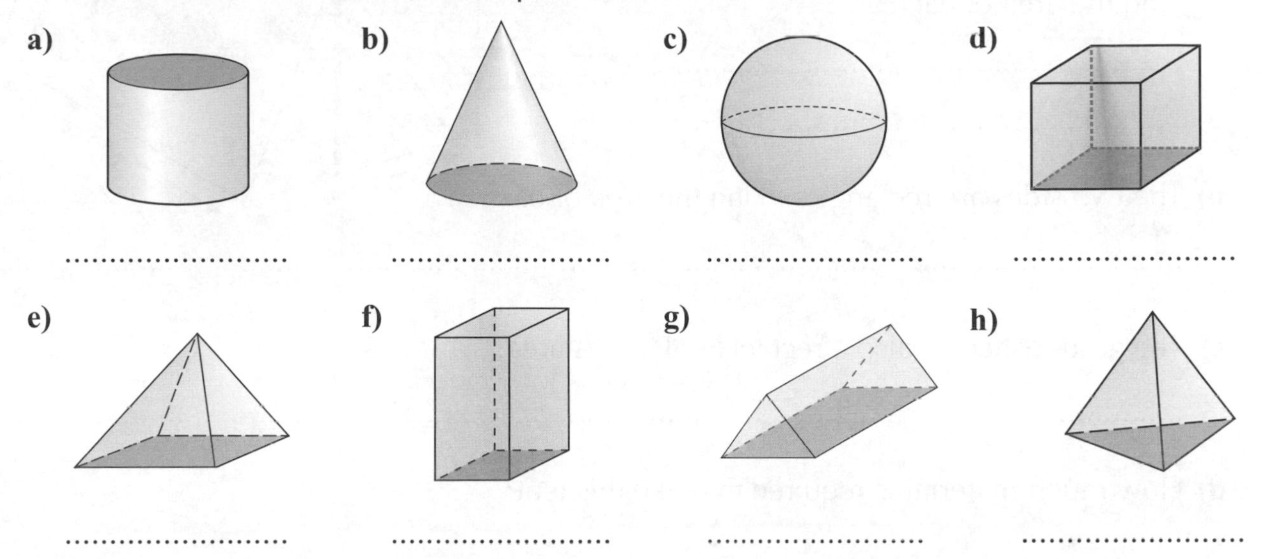

a) .... **b)** .... **c)** .... **d)** ....

e) .... **f)** .... **g)** .... **h)** ....

Q2 State how many of the following the shape on the right has:

a) vertices ....

b) faces ....

c) edges ....

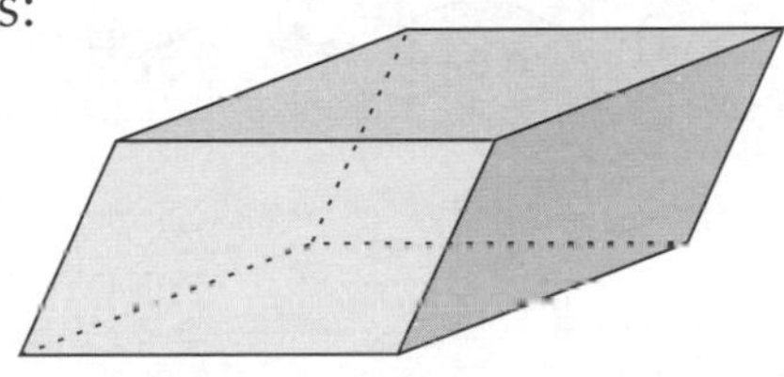

Surface area — just like 'normal' area but in 3D. For some easier shapes, you don't need a formula — you can just add up the areas of the faces.

Q3 Find the surface area of each of the following solids.

a) ....

....

....

....

b) ....

....

....

....

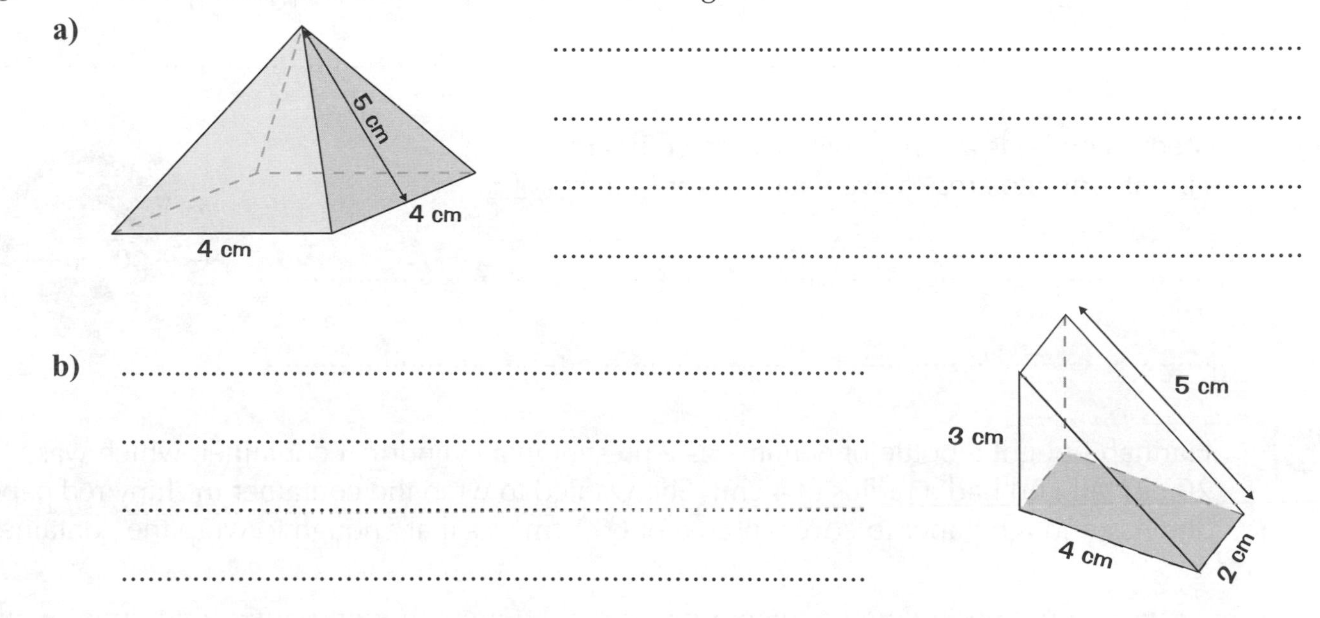

3D Shapes — Surface Area

Q4 A simple tent is to be made in the shape of a triangular prism. The dimensions are shown in the diagram.

3.2 m
3.4 m
4 m
2.3 m

a) The two end-faces are isosceles triangles. Find the area of each.

....

....

b) The two sides are rectangles. Find the area of each.

....

c) The groundsheet is also a rectangle. Find its area.

....

d) How much material is required to make this tent?

....

Q5 Find the surface area of the objects below. Give your answers to 2 decimal places.

a) a sphere with radius 6 cm

6 cm

Surface area of a sphere = $4\pi r^2$

....

b) a cone with radius 3 cm and slant height 5 cm

Surface area of a cone = (πr × slant height) + πr^2

5 cm
3 cm

....

c) a cylinder with radius 8 cm and height 7 cm

8 cm
7 cm

Surface area of a cylinder = $2\pi rh + 2\pi r^2$

....

Q6 Meera's globe is a sphere with diameter 30 cm. Find its surface area. Give your answer <u>in terms of π</u>.

30 cm

....

....

Q7 Hannah bought a bottle of perfume as a present in a cylindrical container, which was 20 cm tall and had a radius of 4 cm. She wanted to wrap the container in shiny red paper. She has enough paper to cover an area of 600 cm². Is that enough to wrap the container?

....

3D Shapes — Volume

Finding volumes of cubes and cuboids is just like finding areas of squares and rectangles — except you've got an extra side to multiply by.

Q1 A match box measures 7 cm by 4 cm by 5 cm.
What is its volume?

....

Q2 What is the volume of a cube with sides of length:

a) 5 cm? ....

b) 9 m? ....

c) 15 mm? ....

Q3 A rectangular swimming pool is 12 m wide and 18 m long.
Its depth is the same throughout. How many cubic metres of water are needed to fill the pool to a depth of 120 cm?

....

....

Q4 An ice cube measures 2 cm by 2 cm by 2 cm.

a) What is its volume? ....

b) Is there enough room in a container measuring 8 cm by 12 cm by 10 cm for 100 ice cubes? ....

Q5

Remember, the volume of <u>any</u> prism is cross-sectional area × length.

Jill buys a bookshelf with the dimensions shown.

a) Find the cross-sectional area.

....

b) Find the volume of the bookshelf in cm^3.

....

Q6 A right-angled triangular prism has the dimensions shown.
Find the volume of the prism in cm^3.

....

....

3D Shapes — Volume

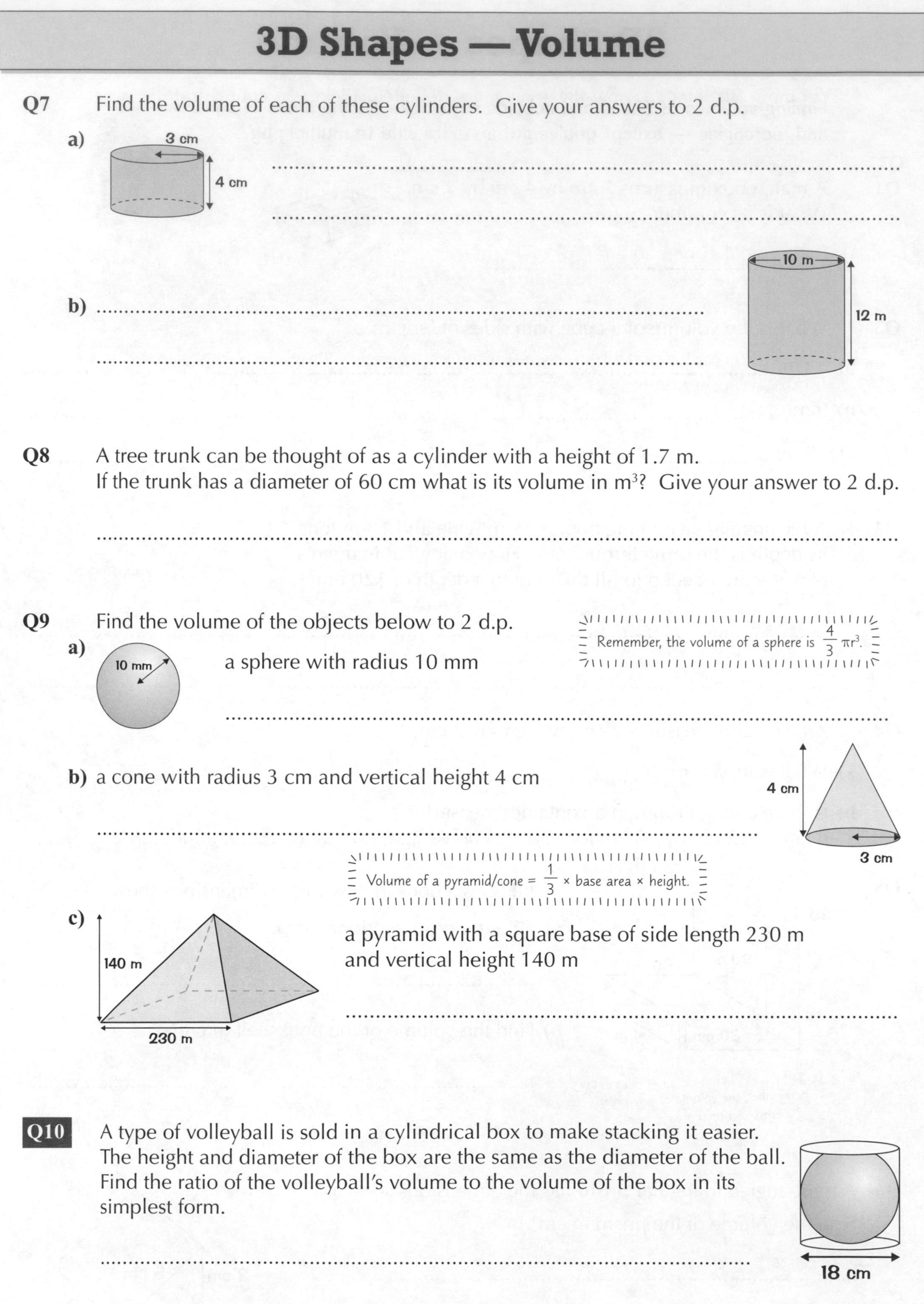

Q7 Find the volume of each of these cylinders. Give your answers to 2 d.p.

a) ....

....

b) ....

....

Q8 A tree trunk can be thought of as a cylinder with a height of 1.7 m.
If the trunk has a diameter of 60 cm what is its volume in m^3? Give your answer to 2 d.p.

....

Q9 Find the volume of the objects below to 2 d.p.

Remember, the volume of a sphere is $\frac{4}{3}\pi r^3$.

a) a sphere with radius 10 mm

....

b) a cone with radius 3 cm and vertical height 4 cm

....

Volume of a pyramid/cone = $\frac{1}{3}$ × base area × height.

c) a pyramid with a square base of side length 230 m
and vertical height 140 m

....

Q10 A type of volleyball is sold in a cylindrical box to make stacking it easier.
The height and diameter of the box are the same as the diameter of the ball.
Find the ratio of the volleyball's volume to the volume of the box in its simplest form.

....

....

3D Shapes — Volume

Yet more 3D things... now these ones are definitely tricky, so concentrate.

Q11 For this question, give your answers in terms of π.
A frustum is made by taking a cone that's 6 cm tall and cutting the top off 2 cm below its peak, as shown on the right.

2 cm
6 cm
1 cm
3 cm

a) The original cone had a radius of 3 cm.
What was the volume of the original cone?

....

b) If the radius of the part of the cone that was cut off is 1 cm, find the volume removed.

....

c) Find the volume of the frustum.

....

Q12 Water is poured into the paddling pool below at a rate of 50 000 cm³ per minute.

a) Calculate the volume of water needed to fill the pool to a depth of 25 cm.

....

b) How long (in minutes) will it take to fill the pool to this depth?

....

....

50 cm
200 cm
150 cm

Q13 A coffee mug is a cylinder closed at one end.
The internal radius is 7 cm and the internal height is 9 cm.

a) Taking π to be 3.14, find the volume of liquid the mug can hold.

....

b) If 1200 cm³ of liquid is poured into the mug, find the depth to the nearest whole cm.

....

....

....

The depth is just the height of the mug taken up by the liquid — which you find by rearranging the volume formula.

Projections

Projections are just different views of a 3D shape.

Q1 The diagram below shows a triangular prism.
On the grid, draw its:

a) front elevation (from the direction of the arrow)
b) side elevation
c) plan view

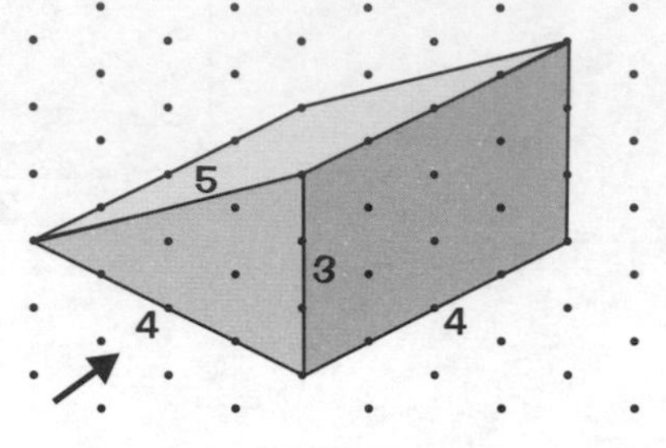

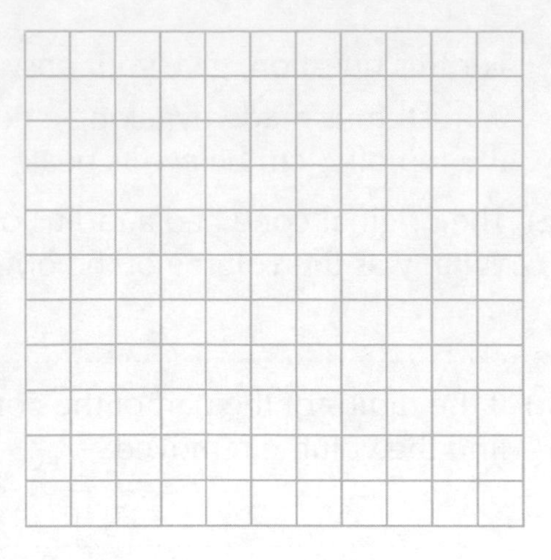

Q2 The projections of a prism made from centimetre cubes are shown below.
How many cubes are needed to build the object?

Q3 The shape below is not drawn to scale.
Draw the front elevation, side elevation and plan view on the grid on the right.

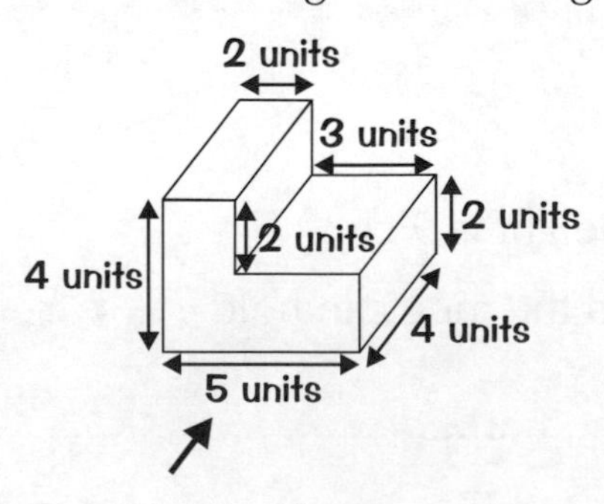

Q4 Use the front elevation, side elevation and plan view of this object to sketch it in 3D.

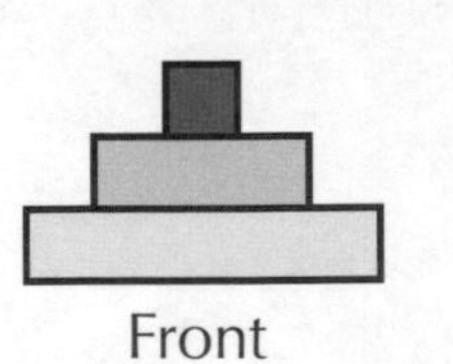
Front

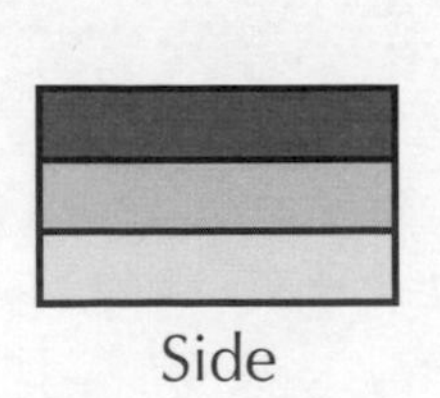
Side

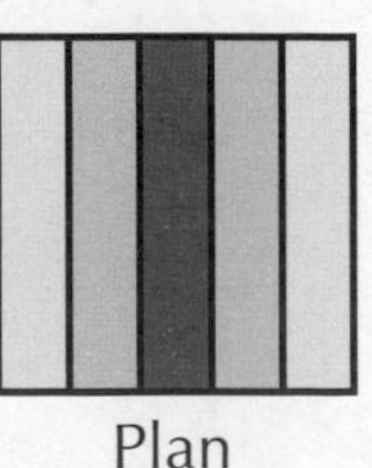
Plan

Angle Basics

Hope you've learnt those angle rules for a straight line and round a point. You'd better learn those fancy angle names too...

Q1 For each of the diagrams below, write down the size of the angle shown.

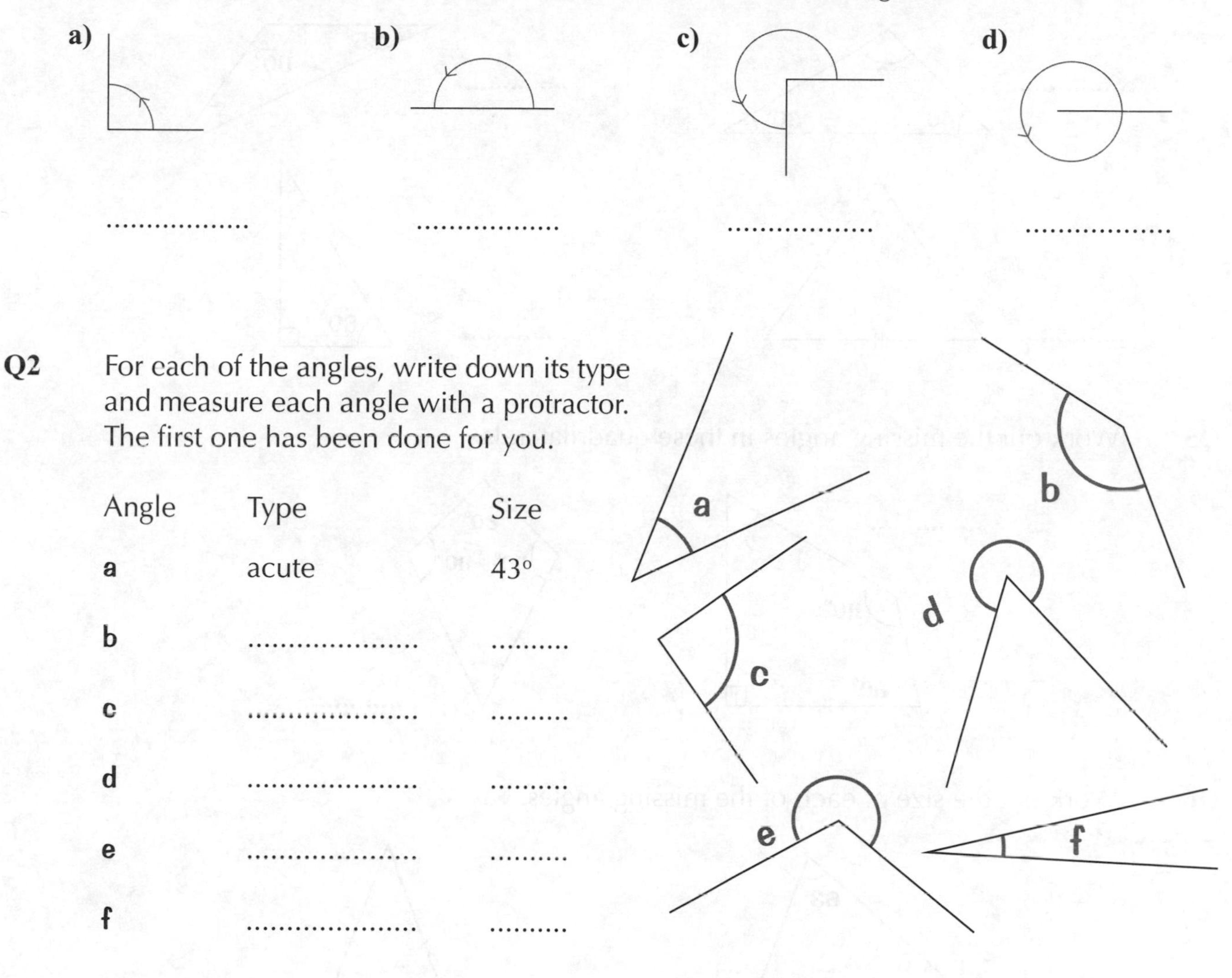

a) .... b) .... c) .... d) ....

Q2 For each of the angles, write down its type and measure each angle with a protractor. The first one has been done for you.

Angle	Type	Size
a	acute	43°
b	....	....
c	....	....
d	....	....
e	....	....
f	....	....

Q3 Work out the angles labelled:

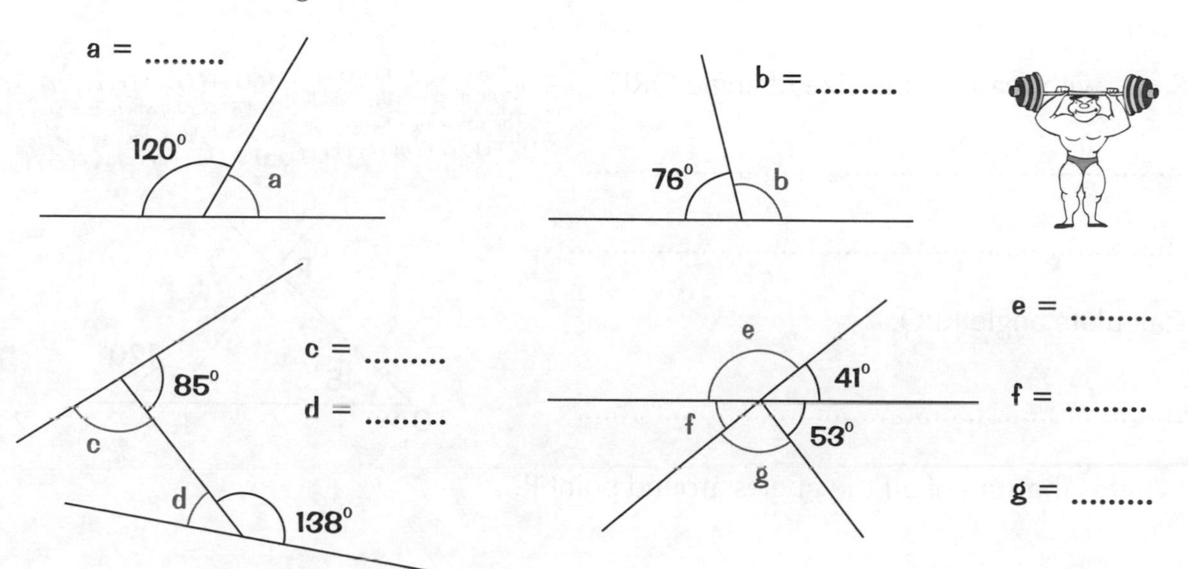

Angle Basics

Those rules for angles in triangles and quadrilaterals aren't that hard, and you'll be well and truly stumped without them. All the more reason to learn them then...

Q4 Work out the missing angle in each of these triangles.
The angles are not drawn to scale so you cannot measure them.

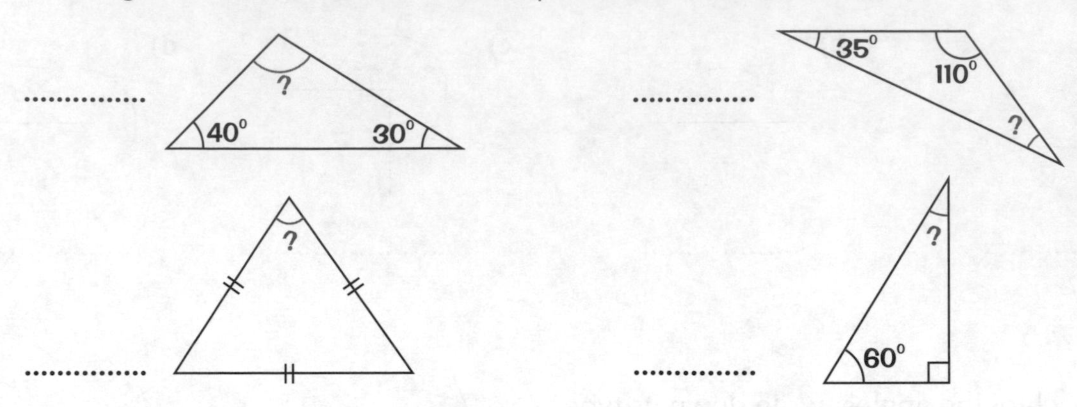

....

....

Q5 Work out the missing angles in these quadrilaterals.

....

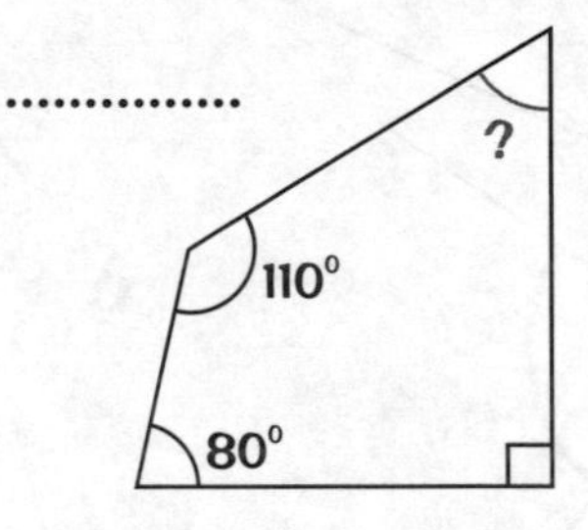

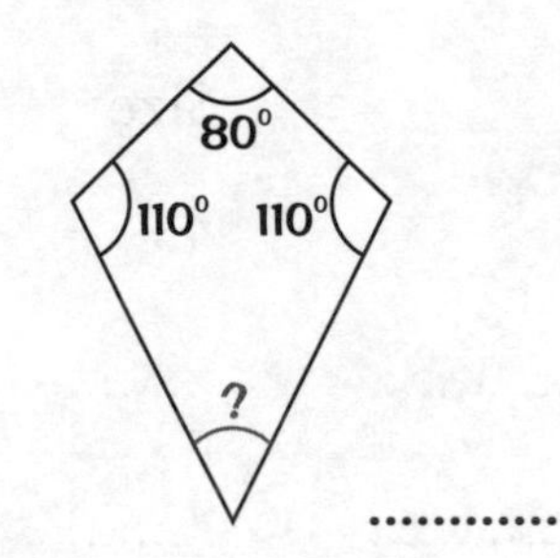

....

Q6 Work out the size of each of the missing angles.

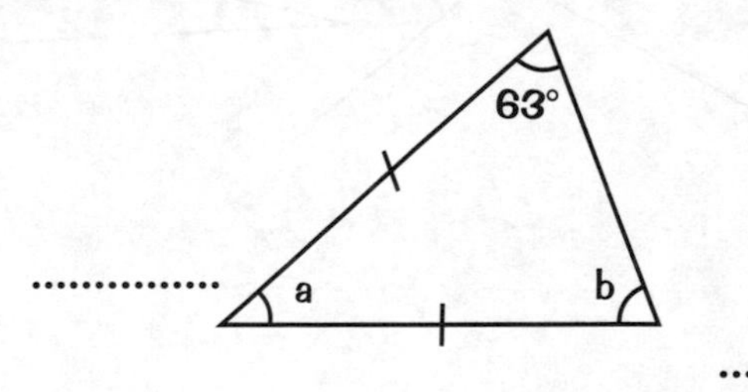

....

....

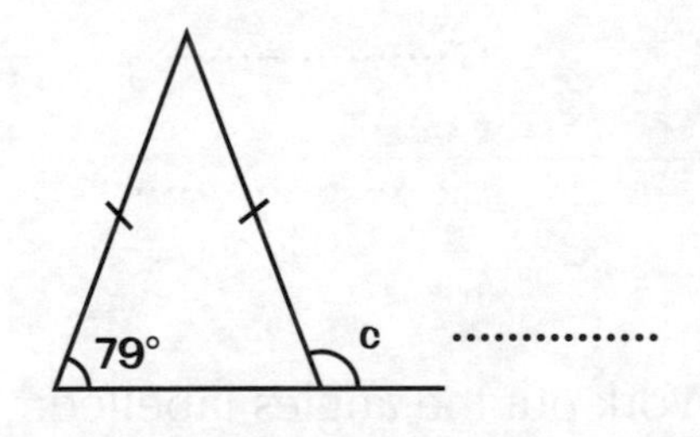

....

Q7 a) State, with reasons, the size of angle QRP.

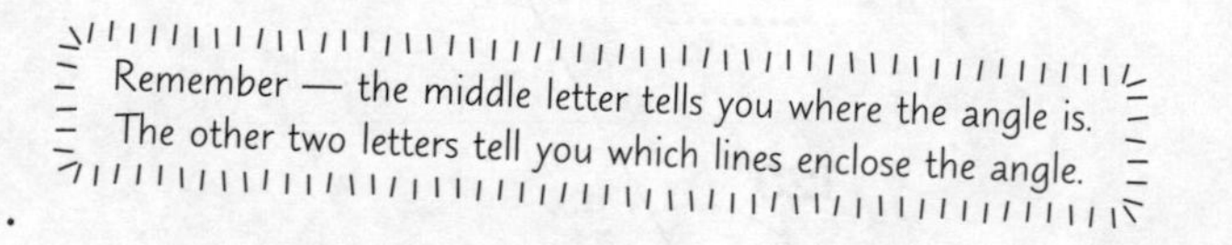

....

....

b) Calculate angle RPQ.

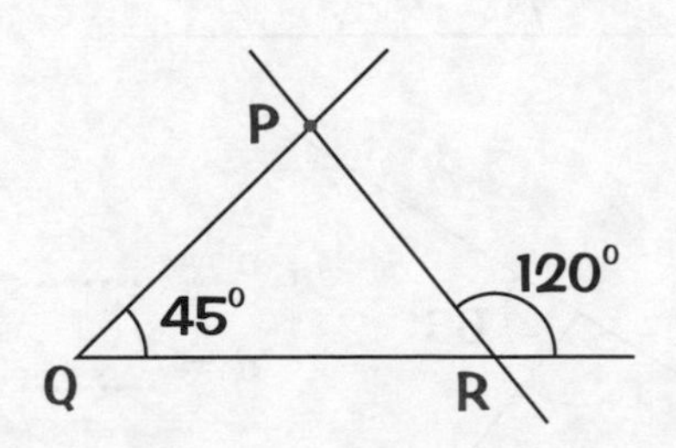

....

c) What is the sum of all the angles around point P?

....

Parallel Lines

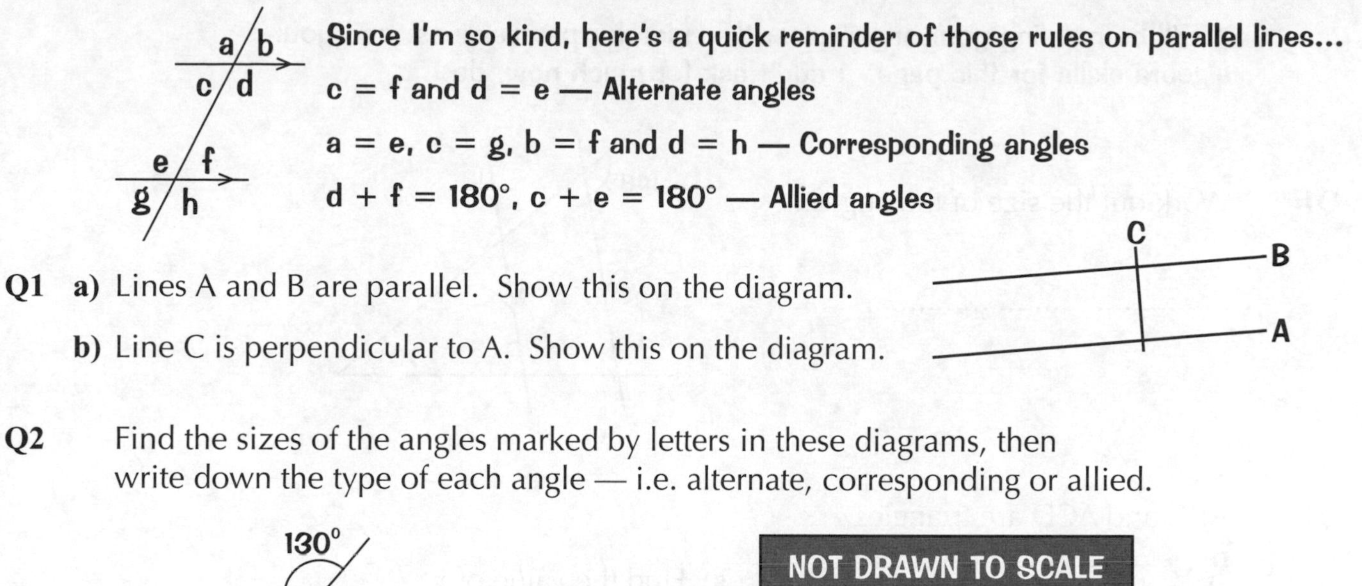

Since I'm so kind, here's a quick reminder of those rules on parallel lines...

c = f and d = e — Alternate angles

a = e, c = g, b = f and d = h — Corresponding angles

d + f = 180°, c + e = 180° — Allied angles

Q1 **a)** Lines A and B are parallel. Show this on the diagram.

b) Line C is perpendicular to A. Show this on the diagram.

Q2 Find the sizes of the angles marked by letters in these diagrams, then write down the type of each angle — i.e. alternate, corresponding or allied.

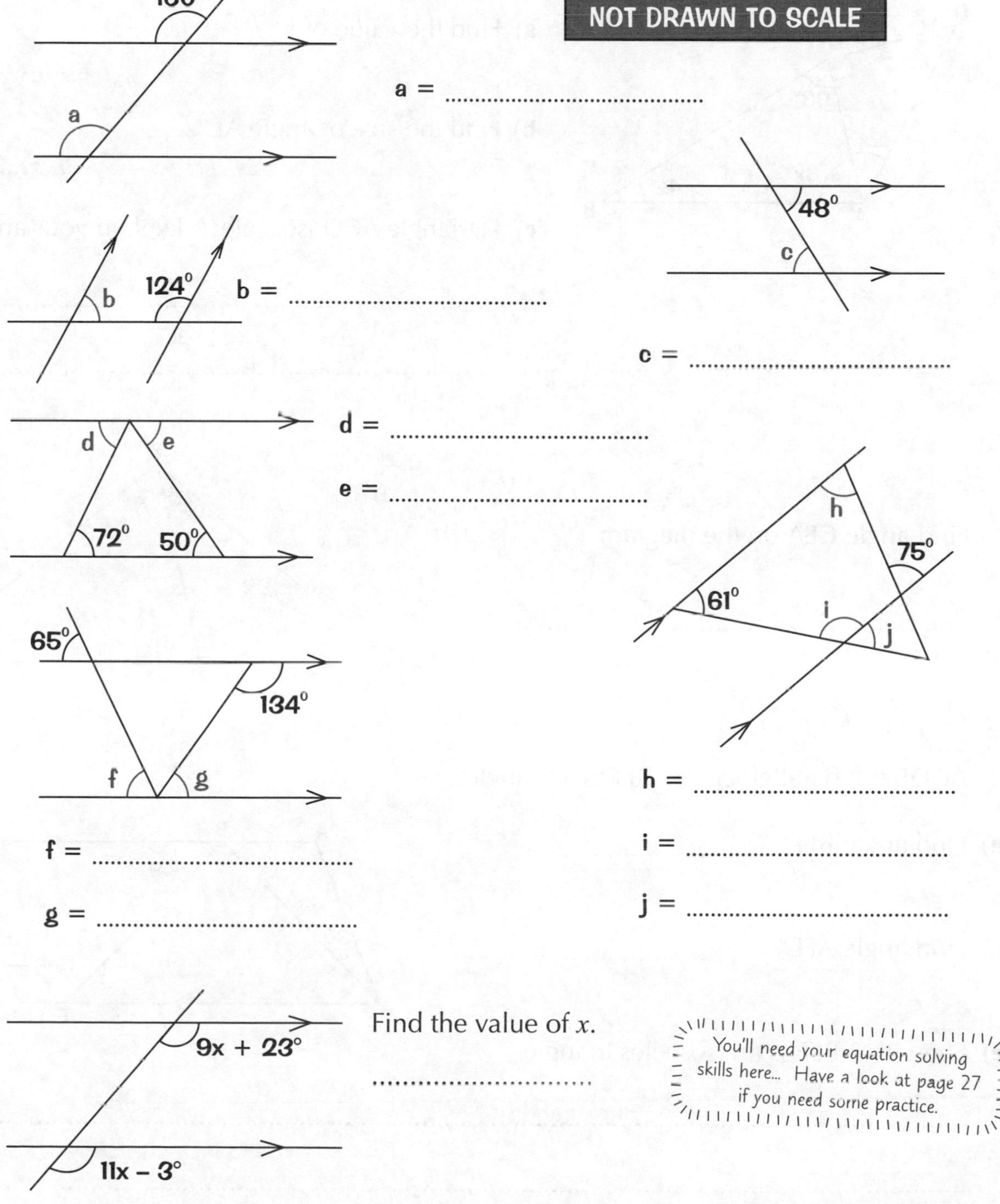

a = ...

b = ...

c = ...

d = ...

e = ...

f = ...

g = ...

h = ...

i = ...

j = ...

Q3 Find the value of x.

....

You'll need your equation solving skills here... Have a look at page 27 if you need some practice.

Geometry Problems

You'll be needing your angle rules, the rules for parallel lines, and your algebra skills for this page. I don't ask for much now, do I...

Q1 Work out the size of the angle x.

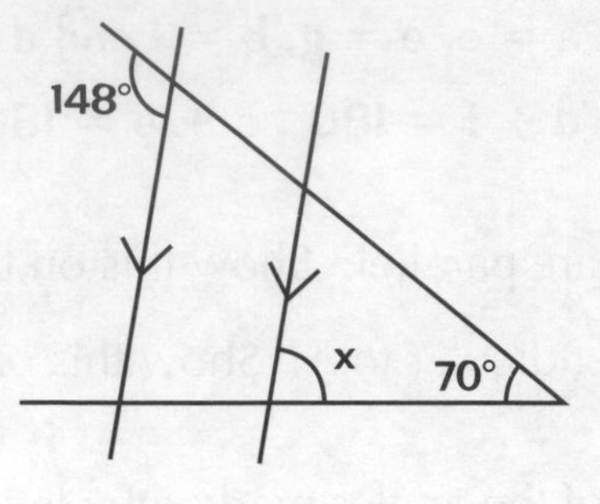

....

Q2 ABC and ACD are triangles.

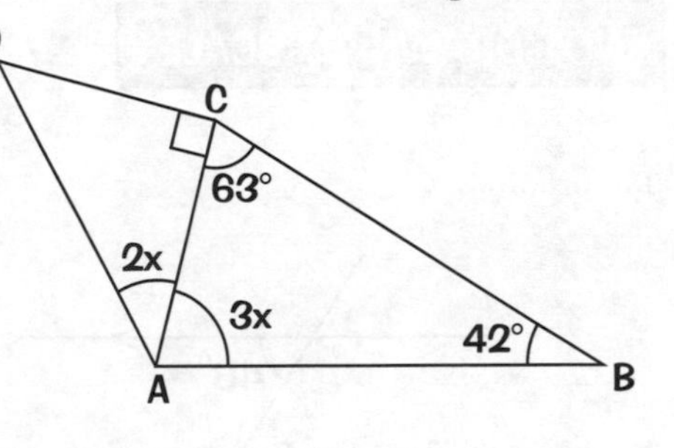

a) Find the value of x. ....

b) Find the size of angle ADC. ....

c) Is triangle ACD isosceles? Explain your answer.

....

....

Q3 Find angle CEA on the diagram.

Find the value of x first.

....

Q4 ACDF is a parallelogram. BEF is a triangle.

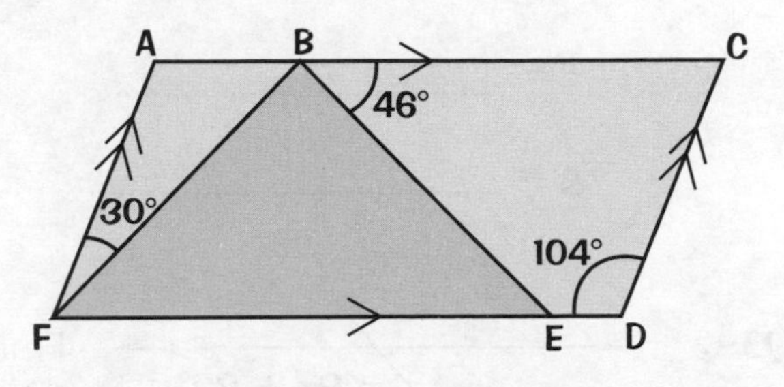

a) Find angle BEF. ....

b) Find angle AFD. ....

c) Show that BEF is an isosceles triangle.

....

....

Angles in Polygons

These three handy formulas work for all polygons (regular or irregular):

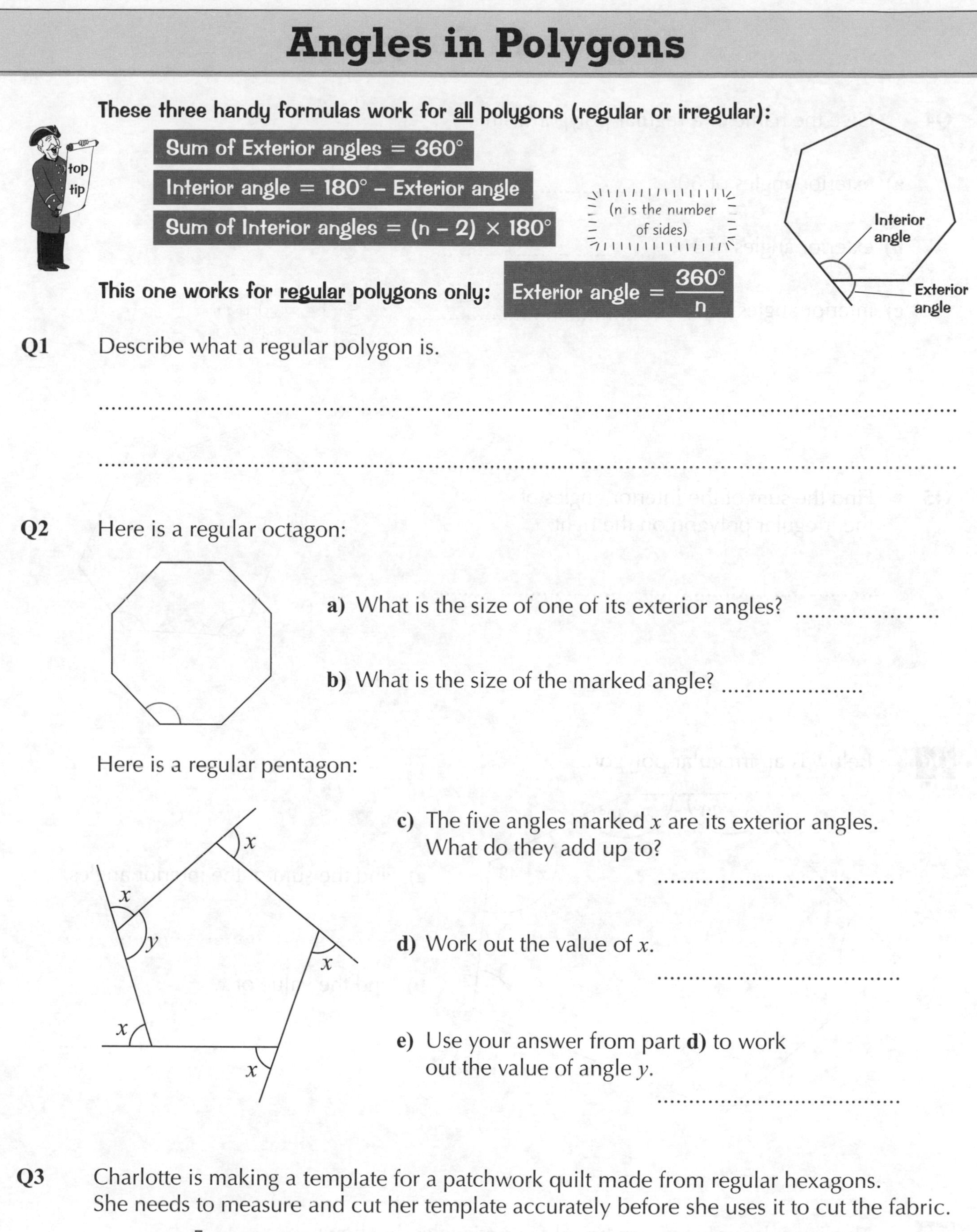

Sum of Exterior angles = 360°

Interior angle = 180° – Exterior angle

Sum of Interior angles = (n – 2) × 180°

(n is the number of sides)

This one works for regular polygons only: Exterior angle = $\frac{360°}{n}$

Q1 Describe what a regular polygon is.

....

....

Q2 Here is a regular octagon:

a) What is the size of one of its exterior angles? ....

b) What is the size of the marked angle? ....

Here is a regular pentagon:

c) The five angles marked x are its exterior angles. What do they add up to?

....

d) Work out the value of x.

....

e) Use your answer from part **d)** to work out the value of angle y.

....

Q3 Charlotte is making a template for a patchwork quilt made from regular hexagons. She needs to measure and cut her template accurately before she uses it to cut the fabric.

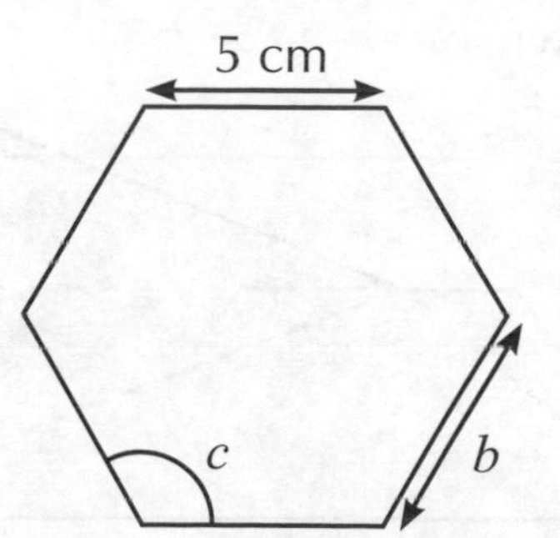

a) How long should she make side b? ....

b) What is the size of angle c? ....

Angles in Polygons

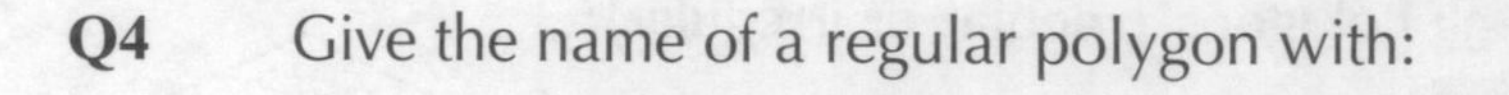

Q4 Give the name of a regular polygon with:

a) exterior angles of 60° ....

b) exterior angles of 36° ....

c) interior angles of 140° ....

For part c), find the exterior angle size first.

Q5 Find the sum of the interior angles of the irregular polygon on the right.

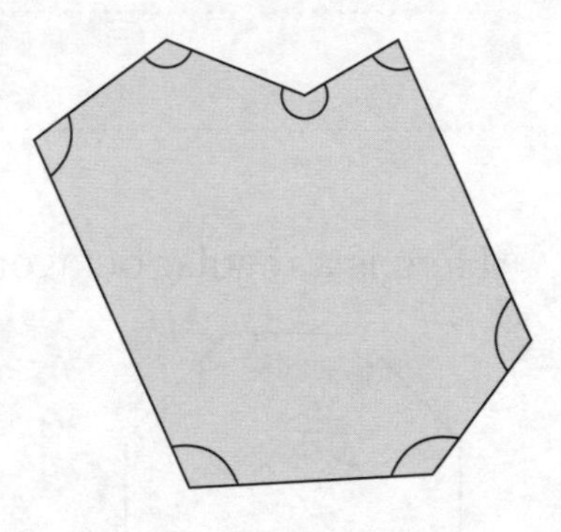

....

Q6 Below is an irregular polygon.

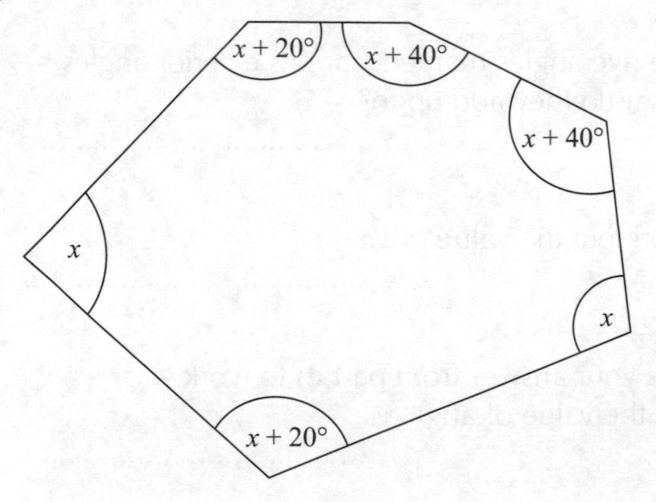

a) Find the sum of the interior angles.

....

b) Find the value of x.

....

Q7 This irregular polygon can be split into triangles, as shown.

By considering the angles in the triangles, show that the sum of the interior angles of the polygon is 720°.

....

....

Loci and Construction

You've gotta be precise with these — always use a ruler and compasses. Make sure you leave in the construction marks — they might gain you a few extra marks.

Q1 Make an accurate drawing below of the triangle on the right.

Measure side AB on your triangle, giving your answer in millimetres.

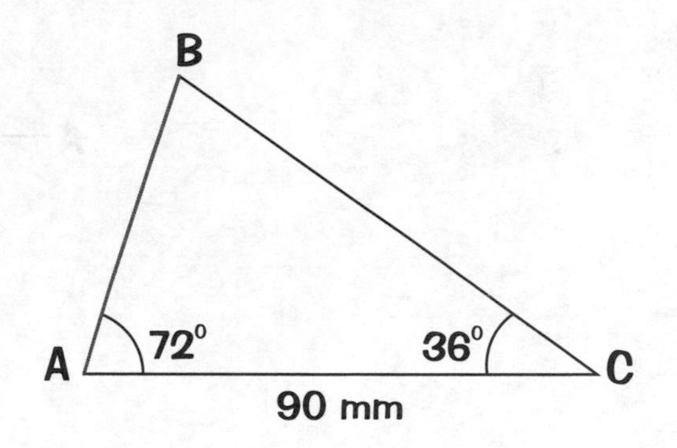

AB = ... mm

Q2 a) In the space to the right, construct a triangle ABC with AB = 4 cm, BC = 5 cm, AC = 3 cm.

b) Construct the perpendicular bisector of AB and where this line meets BC, label the new point D.

Remember — a <u>bisector</u> is a line splitting an angle or line exactly in two.

Q3

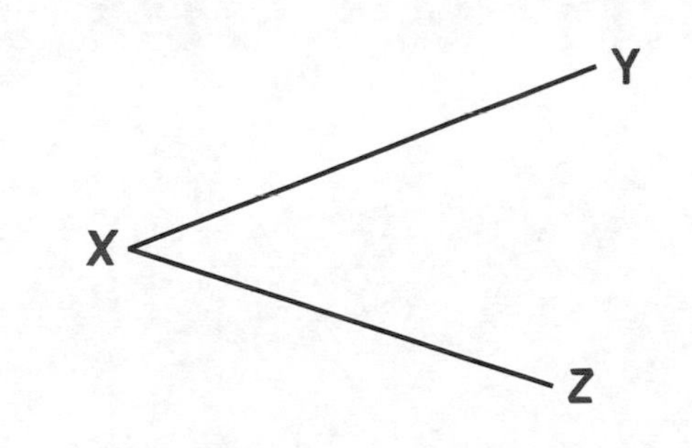

The diagram on the left shows two lines XY and XZ which meet at the point X. Construct the angle bisector of YXZ.

Don't forget — a <u>locus</u> is a line or region showing all the points that fit a given rule.

Q4 On a plain piece of paper mark two points A and B which are 6 cm apart.

a) Draw the locus of points which are 4 cm from A.

b) Draw the locus of points which are 3 cm from B.

c) There are 2 points which are both 4 cm from A and 3 cm from B. Label them X and Y.

Loci and Construction

Q5 Use a ruler and compasses to draw the perpendicular from the point C to the line AB.

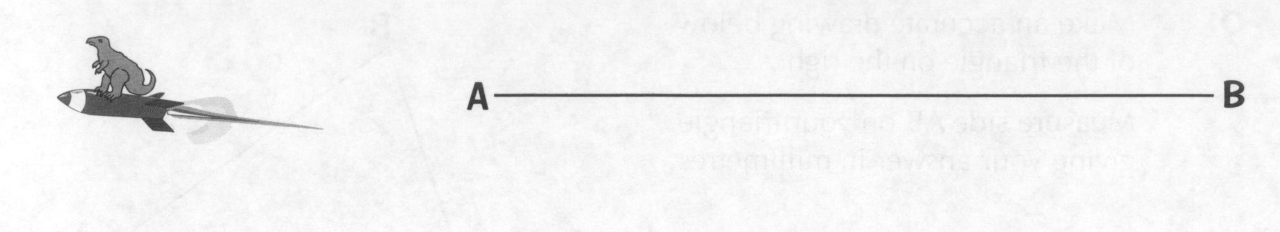

C

Q6 Penny is rearranging her living room. She has started to draw a diagram, shown below, to plan where she will put her furniture. On the diagram, the rest of Penny's furniture should be at least 3 cm away from her coffee table on all sides. Using a ruler and compasses, show the region where she cannot place her furniture.

coffee table

Q7 With the aid of a pair of compasses, accurately draw an equilateral triangle with sides 5 cm.

Loci and Construction

Q8 Using a pair of compasses, accurately draw a square with sides 6 cm.

Q9 A planner is trying to decide where to build a house. He is using the plan below, which shows the location of a canal, a TV mast **(A)** and a water main **(B)**.
The house must be more than 100 m away from both the canal and the TV mast and be within 60 m of the water main.
Shade the area on the map where the planner could build the house.

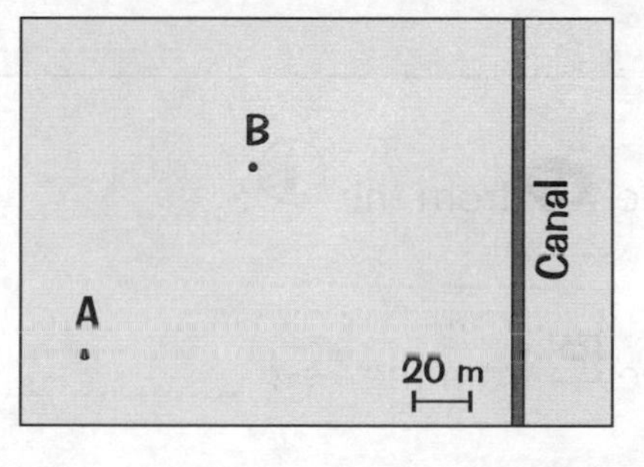

Q10 The right-angled triangle ABC below is rotated 90° clockwise about vertex A.
Draw the locus of vertex B.

Don't panic with this one — rotate the whole triangle and think about how B is moving...

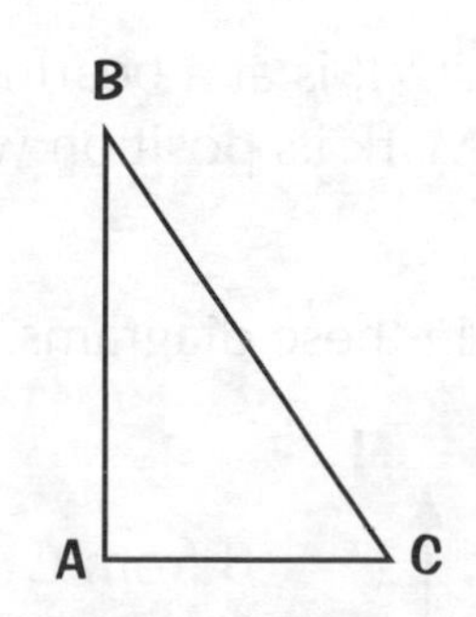

Bearings

Bearings always have three digits — even the small ones... in other words, if you've measured 60°, you've got to write it as 060°.

This is a map of part of a coastline.

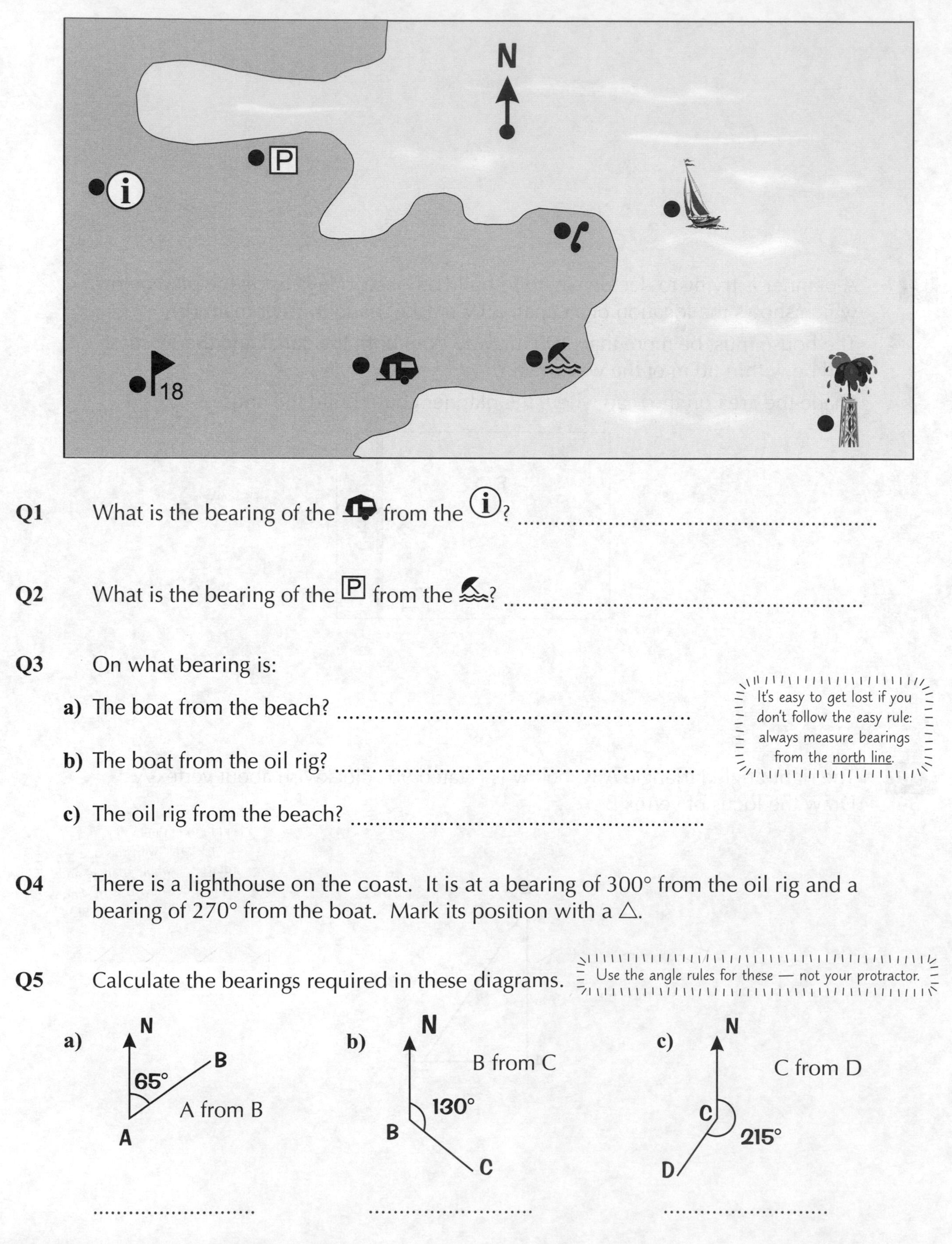

Q1 What is the bearing of the [caravan] from the (i)? ....

Q2 What is the bearing of the [P] from the [beach]? ....

Q3 On what bearing is:

a) The boat from the beach? ....

b) The boat from the oil rig? ....

c) The oil rig from the beach? ....

It's easy to get lost if you don't follow the easy rule: always measure bearings from the north line.

Q4 There is a lighthouse on the coast. It is at a bearing of 300° from the oil rig and a bearing of 270° from the boat. Mark its position with a △.

Q5 Calculate the bearings required in these diagrams.

Use the angle rules for these — not your protractor.

a)

....

b)

....

c)

....

Maps and Scale Drawings

If the scale doesn't say what units it's in, it just means that both sides of the ratio are the same units — so 1:1000 would mean 1 cm:1000 cm.

Q1 The scale on this map is 1 cm:4 km.

a) Measure the distance from A to B in cm. ....

b) What is the actual distance from A to B in km? ....

c) A helicopter flies on a direct route from A to B, B to C and C to D.
What is the total distance flown in km?

....

d) Angus walks 8 km from point B on a bearing of 130°.
Mark his new position on the map and label it E.

Q2 Frank has made a scale drawing of his garden to help him plan some improvements.
The scale on the drawing is 1:50.

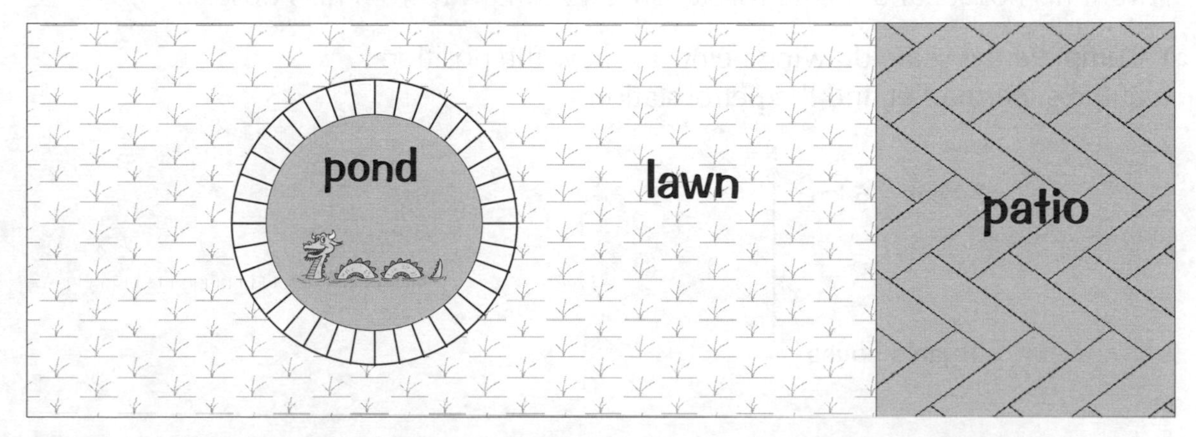

a) Frank wants to put up a fence along the three outside edges of the lawn.
How many metres of fencing does he need to buy?

....

b) What are the actual dimensions of Frank's patio in m?

....

Maps and Scale Drawings

Watch out for those units... there's quite a mixture here — you'll have to convert some of them before you can go anywhere.

Q3 A rectangular room measures 20 m long and 15 m wide. Work out the measurements for a scale drawing of the room using a scale of 1 cm = 2 m.

Length = ... Width = ...

Q4 A rectangular field is 60 m long and 40 m wide. The farmer needs to make a scale drawing of it. He uses a scale of 1 : 2000. Work out the measurements for the scale drawing. (Hint — change the m to cm.)

....

Q5 A rectangular room is 4.8 m long and 3.6 m wide. Make a scale drawing of it using a scale of 1 cm to 120 cm. First work out the measurements for the scale drawing.

Length = ...

Width = ...

On your scale drawing mark a window, whose actual length is 2.4 m, on one long wall and mark a door, actual width 90 cm, on one short wall.

Window = ...

Door = ...

Q6 There is a supermarket on a bearing of 115° from Ryan's house and a petrol station on a bearing of 085° from his house. The supermarket is 20 km away from his house, and the petrol station is 15 km away from his house.

a) Complete the scale drawing below to show the position of the supermarket and the petrol station.

N

1 cm = 5 km

Ryan's house

b) How far is the petrol station from the supermarket in real life, to the nearest km?

....

Pythagoras' Theorem

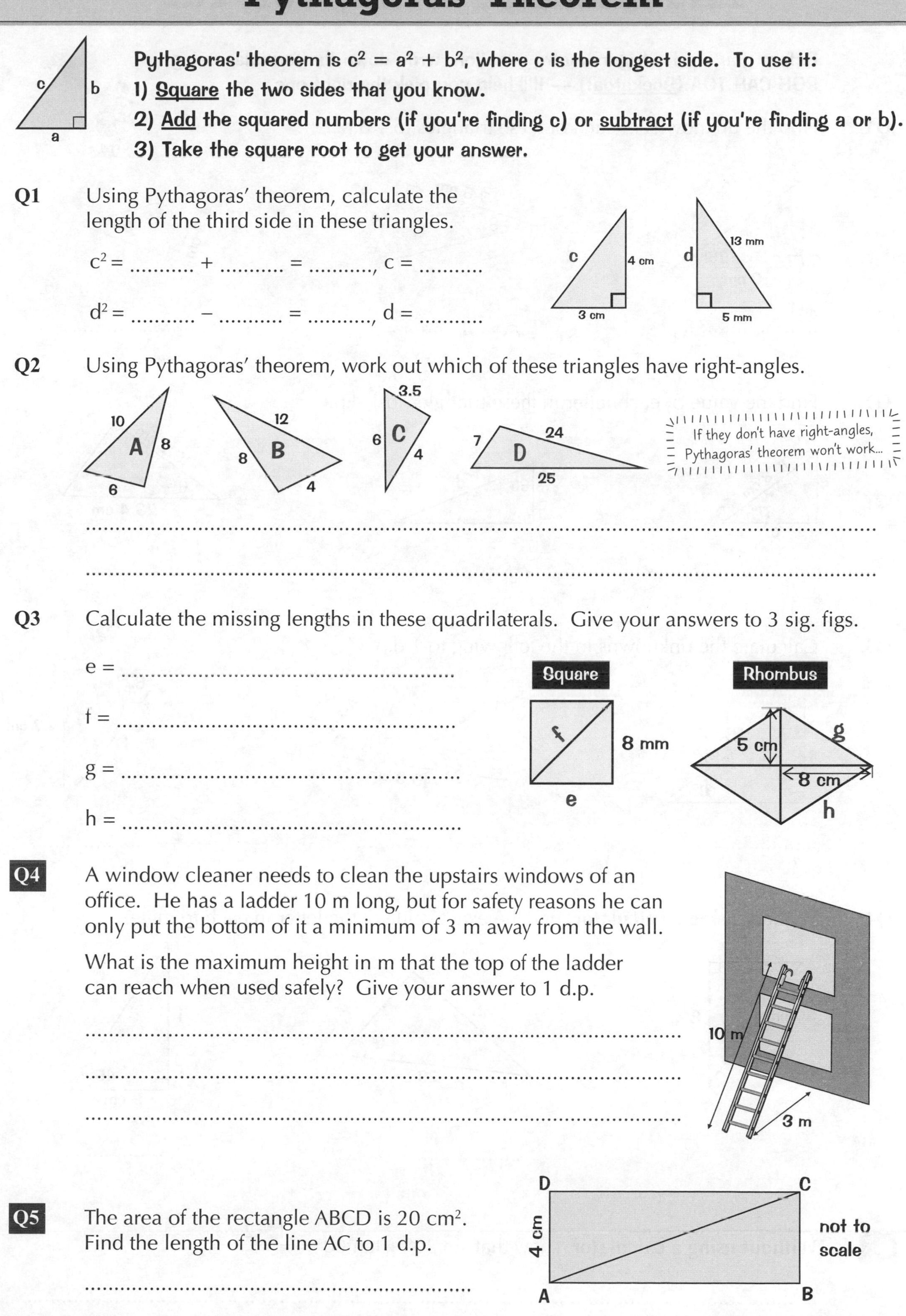

Pythagoras' theorem is $c^2 = a^2 + b^2$, where c is the longest side. To use it:

1) Square the two sides that you know.

2) Add the squared numbers (if you're finding c) or subtract (if you're finding a or b).

3) Take the square root to get your answer.

Q1 Using Pythagoras' theorem, calculate the length of the third side in these triangles.

c^2 = ... + ... = ..., c = ...

d^2 = ... − .... = ..., d = ...

Q2 Using Pythagoras' theorem, work out which of these triangles have right-angles.

....

....

Q3 Calculate the missing lengths in these quadrilaterals. Give your answers to 3 sig. figs.

e = ...

f = ...

g = ...

h = ...

Q4 A window cleaner needs to clean the upstairs windows of an office. He has a ladder 10 m long, but for safety reasons he can only put the bottom of it a minimum of 3 m away from the wall.

What is the maximum height in m that the top of the ladder can reach when used safely? Give your answer to 1 d.p.

....

....

....

Q5 The area of the rectangle ABCD is 20 cm². Find the length of the line AC to 1 d.p.

....

Trigonometry — Sin, Cos, Tan

Before you start a trigonometry question, write down the formulas using SOH CAH TOA (Sockatoa!) — it'll help you pick the right one.

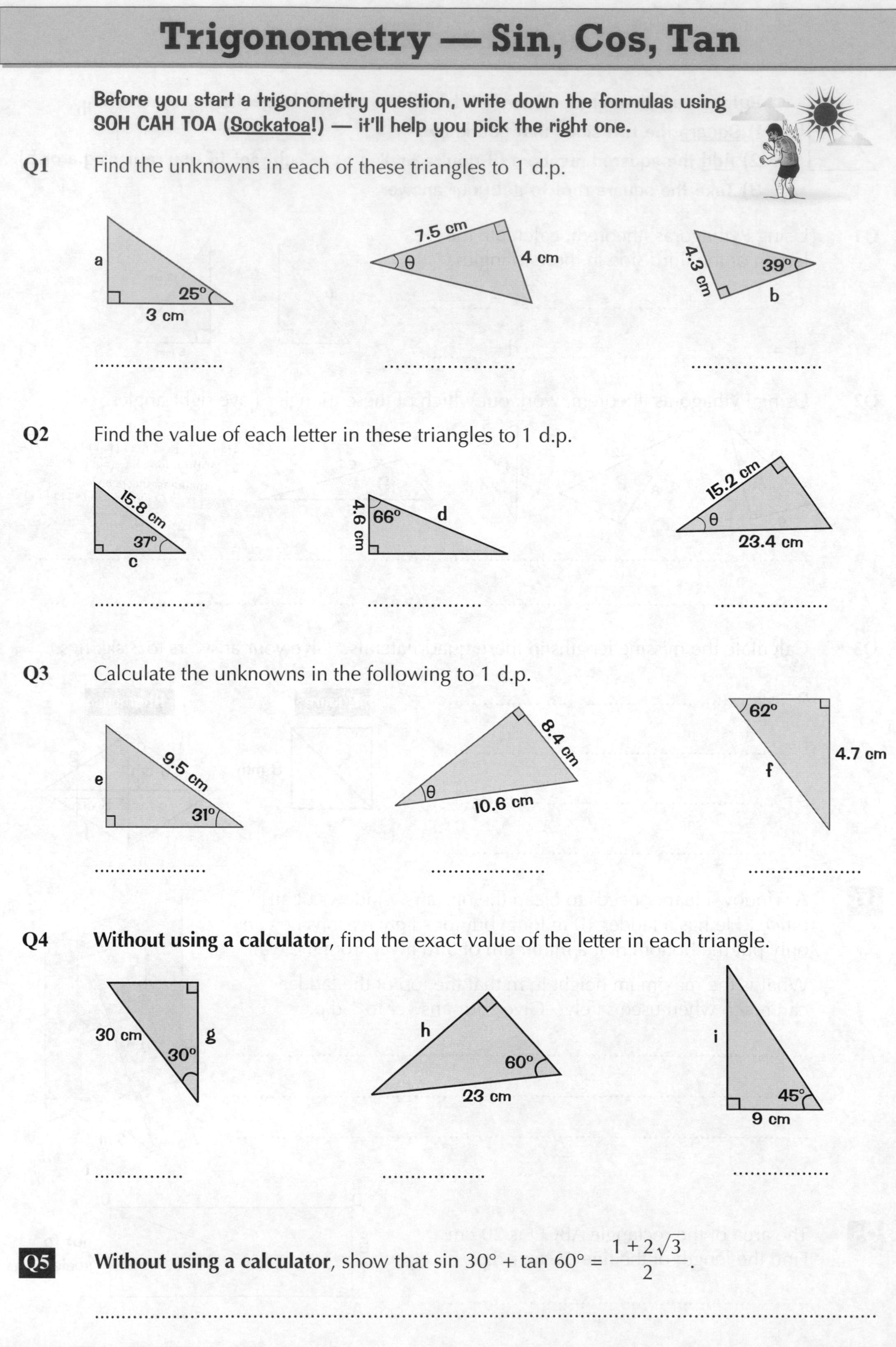

Q1 Find the unknowns in each of these triangles to 1 d.p.

....

Q2 Find the value of each letter in these triangles to 1 d.p.

....

Q3 Calculate the unknowns in the following to 1 d.p.

....

Q4 **Without using a calculator**, find the exact value of the letter in each triangle.

....

Q5 **Without using a calculator**, show that $\sin 30° + \tan 60° = \frac{1 + 2\sqrt{3}}{2}$.

....

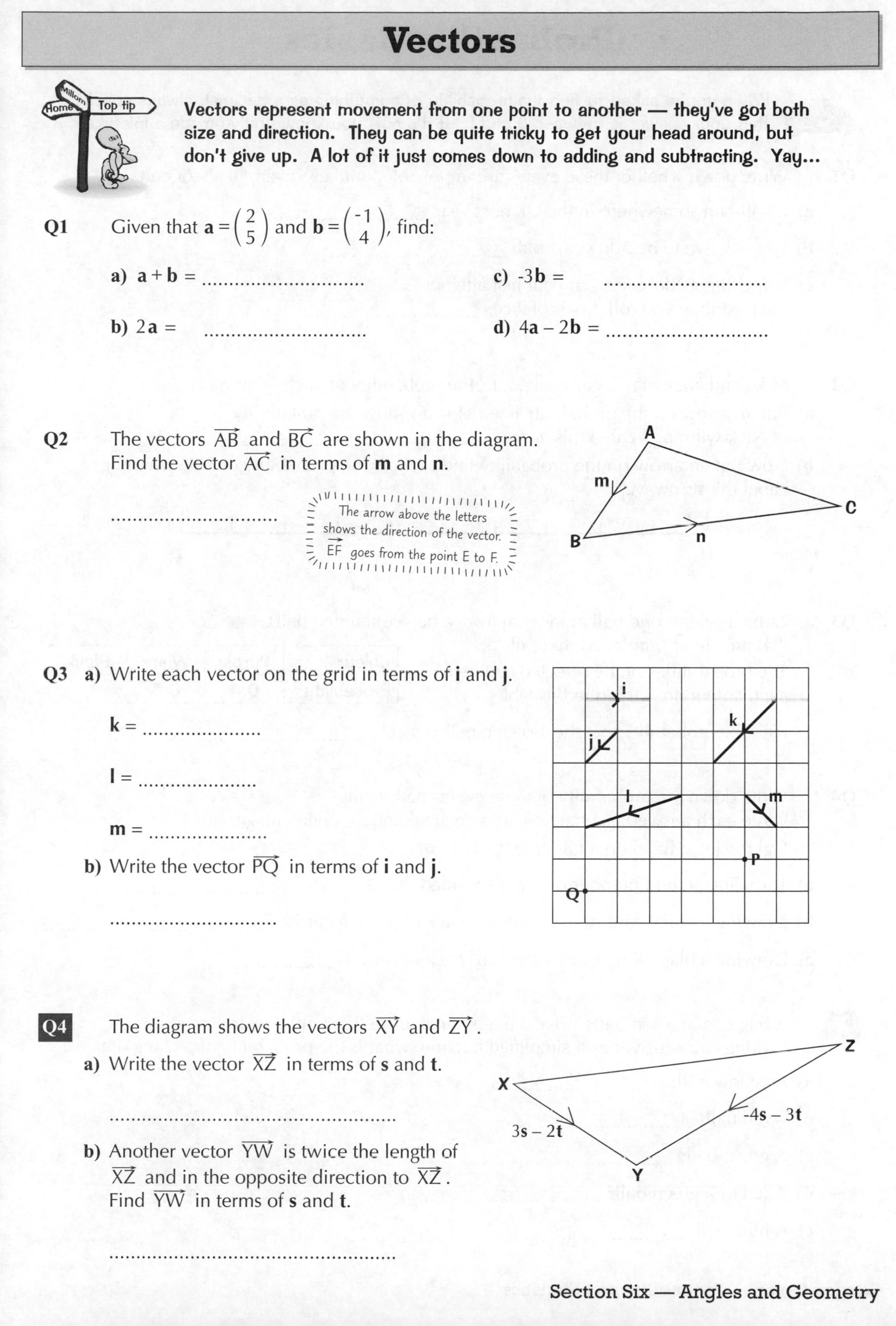

Vectors

Top tip

Vectors represent movement from one point to another — they've got both size and direction. They can be quite tricky to get your head around, but don't give up. A lot of it just comes down to adding and subtracting. Yay...

Q1 Given that $\mathbf{a} = \begin{pmatrix} 2 \\ 5 \end{pmatrix}$ and $\mathbf{b} = \begin{pmatrix} -1 \\ 4 \end{pmatrix}$, find:

a) $\mathbf{a} + \mathbf{b}$ = ...

b) $2\mathbf{a}$ = ...

c) $-3\mathbf{b}$ = ...

d) $4\mathbf{a} - 2\mathbf{b}$ = ...

Q2 The vectors $\overrightarrow{AB}$ and $\overrightarrow{BC}$ are shown in the diagram. Find the vector $\overrightarrow{AC}$ in terms of **m** and **n**.

....

The arrow above the letters shows the direction of the vector. $\overrightarrow{EF}$ goes from the point E to F.

Q3 a) Write each vector on the grid in terms of **i** and **j**.

k = ...

l = ...

m = ...

b) Write the vector $\overrightarrow{PQ}$ in terms of **i** and **j**.

....

Q4 The diagram shows the vectors $\overrightarrow{XY}$ and $\overrightarrow{ZY}$.

a) Write the vector $\overrightarrow{XZ}$ in terms of **s** and **t**.

....

b) Another vector $\overrightarrow{YW}$ is twice the length of $\overrightarrow{XZ}$ and in the opposite direction to $\overrightarrow{XZ}$. Find $\overrightarrow{YW}$ in terms of **s** and **t**.

....

Probability Basics

When you're asked to find the probability of something as a decimal, always check that your answer is between 0 and 1. If it's not, you know you've made a mistake.

Q1 Write down whether these events are impossible, unlikely, even, likely or certain.

a) It will rain somewhere in the UK next year. ....

b) You will live to be 300 years old. ....

c) You will get "double-6" at your first attempt the next time you roll a pair of dice. ....

Q2 Mike and Nick play a game of pool. The probability of Nick winning is $\frac{7}{10}$.

a) Put an arrow on the probability line below to show the probability of Nick winning. Label this arrow N.

b) Now put an arrow on the probability line to show the probability of Mike winning. Label this arrow M.

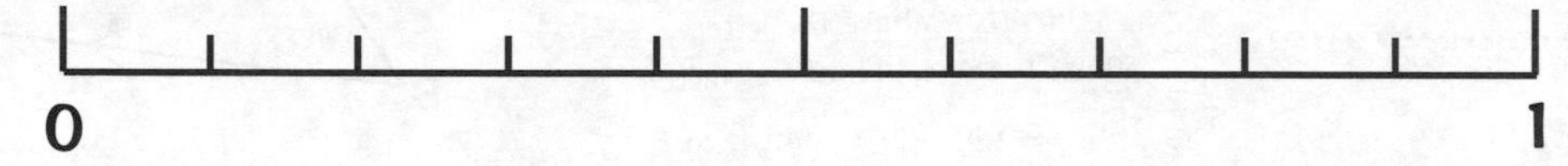

Q3 Tamara selects one ball at random from a bag containing balls that are either purple, white or black. The probabilities of the selected ball being each colour are shown in this table.

Colour	Purple	White	Black
Probability	0.4	0.5	

Find the probability that the selected ball is black. ....

Q4 Write down the probability of these events happening. Write each answer as a fraction, as a decimal and as a percentage. E.g. tossing a head on a fair coin: ½, 0.5, 50%

a) Throwing an odd number on a fair six-sided dice. ...., ..., ...

b) Drawing a black card from a standard pack of playing cards. ...., ..., ...

c) Drawing a black King from a standard pack of cards. ...., ..., ...

Q5 A bag contains ten balls. Five are red, three are yellow and two are green. Writing each answer as a simplified fraction, what is the probability of picking out:

a) A yellow ball? ....

b) A red ball? ....

c) A green ball? ....

d) A red or a green ball? ....

e) A blue ball? ....

Listing Outcomes

Q1 The outcome when a coin is tossed is heads (H) or tails (T). Complete this table of outcomes when two coins are tossed one after the other.

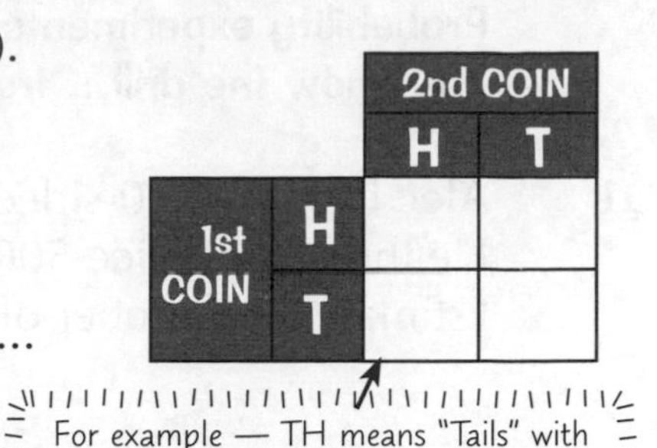

For example — TH means "Tails" with the 1st coin and "Heads" with the 2nd.

a) How many possible outcomes are there? ....

b) What is the probability of getting 2 heads? ....

c) What is the probability of getting a head followed by a tail? ....

In the exam, you might not be asked to put the possible outcomes in a table. But it's a good idea to make your own table anyway — that way you don't miss anything out.

Q2 Lucy has designed a game for the school fair. Two dice are rolled at the same time and the scores on the dice are added together. She needs to work out the probability of all the possible outcomes so she can decide which outcomes will get a prize. Complete the table of possible outcomes below.

	2nd DICE					
1st DICE	1	2	3	4	5	6
1						
2	3					
3						
4						
5			8			
6						

a) How many different ways of rolling the two dice are there? ....

b) What is the probability of scoring:

i) 2 ....

ii) 6 ....

iii) 10 ....

iv) More than 9 ....

v) Less than 4 ....

vi) More than 12 ....

c) Lucy would like to give a prize for an outcome whose probability is exactly 50%. Suggest an outcome whose probability is exactly 50%.

....

Q3 Elsie has the following 3 cards.

a) How many different 3-digit numbers can she make using the cards? ....

b) How many of these numbers are less than 300? ....

c) How many of the numbers are even? ....

Probability Experiments

Probability experiments involve doing things over and over again. You know the drill... try these questions and see how well you do.

Q1 Alec has a fair 10-sided dice with sides numbered 1-10.
He throws the dice 500 times.
Estimate the number of times the dice will land on 3. ....

Q2 150 learner drivers recently took their driving test.
- 100 of these people had taken professional driving lessons.
- 90 of the people who had taken professional driving lessons passed their test.
- 99 people altogether passed their test.

a) Use the information above to complete this frequency tree.

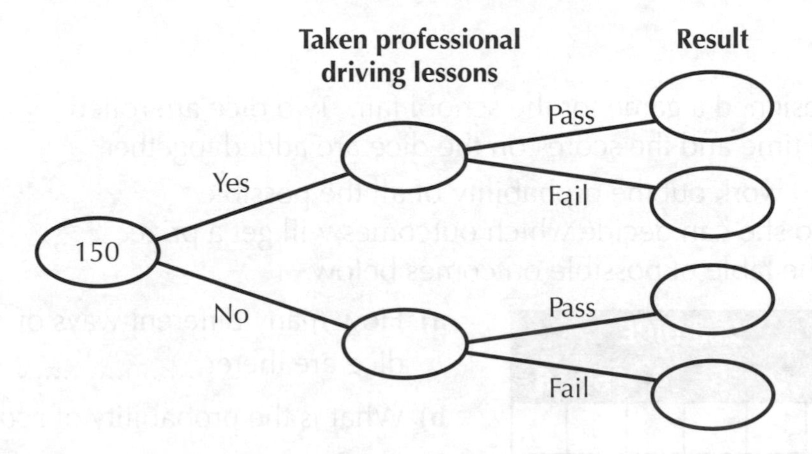

b) A person who did not take professional lessons is selected at random.
Find the probability that they passed their driving test. ....

Q3 Kenny has a machine that he hopes will generate random numbers between 1 and 5, but he thinks the machine has a bug that makes it biased.
To test it, he makes it generate a number 100 times. Here are the results of Kenny's test:

Number	1	2	3	4	5
Number of times generated	18	19	22	21	20
Relative frequency					

a) If the machine is unbiased, what would you expect the relative frequency of each number to be? ....

b) Fill in the third row of the table to show the relative frequency of each number.

c) Do you think the machine is biased? Give a reason for your answer.

....

d) If the machine is unbiased and Kenny repeats the test 100 000 times:

i) What would you expect to happen to the relative frequency of each number?

....

ii) What would be the expected frequency of each of the numbers 1-5? ....

The AND / OR Rules

- **The AND Rule:** The probability of event A <u>AND</u> event B happening is equal to the probability of event A <u>MULTIPLIED BY</u> the probability of event B (for independent events).
- **The OR Rule:** The probability of <u>EITHER</u> event A <u>OR</u> event B happening is equal to the probability of event A <u>ADDED TO</u> the probability of event B (for events that can't happen at the same time).

Q1 The spinner on the right has segments that are coloured either black or white. The probability that the spinner lands on a black segment is 0.6.

a) Find the probability that the spinner lands on a white segment. ....

b) The spinner is spun twice.

i) Find the probability that the spinner lands on black both times. ....

ii) Find the probability that the spinner lands on white on both spins. ....

Q2 Mishal has a bag containing balls that are either red, green, purple or blue. She selects one ball at random from the bag. The probability of the ball being each colour is shown in the table.

Colour	Red	Green	Purple	Blue
Probability	0.25	0.4		0.15

a) Calculate the probability that Mishal selects a purple ball. ....

b) Calculate the probability that Mishal selects either a red ball or a green ball.

....

c) What is the probability that Mishal selects a ball that's red, green or purple?

....

Q3 Janine has an ordinary pack of playing cards. She selects one card at random.

a) What is the probability that Janine selects a red card?

....

b) What is the probability that Janine selects a club?

....

c) Use your answers to find the probability that Janine selects either a red card or a club.

....

Q4 The two spinners on the right are spun at the same time. Find the probability of:

a) Spinning an A and a 2. ....

b) Spinning a 1 and a 3. ....

c) Spinning <u>either</u> A and 1 <u>or</u> C and 3. ....

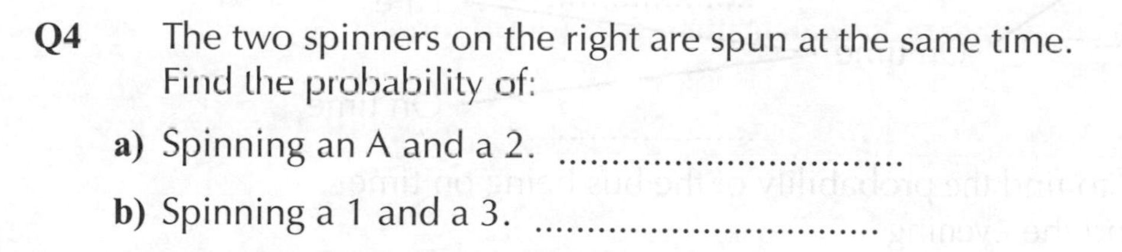

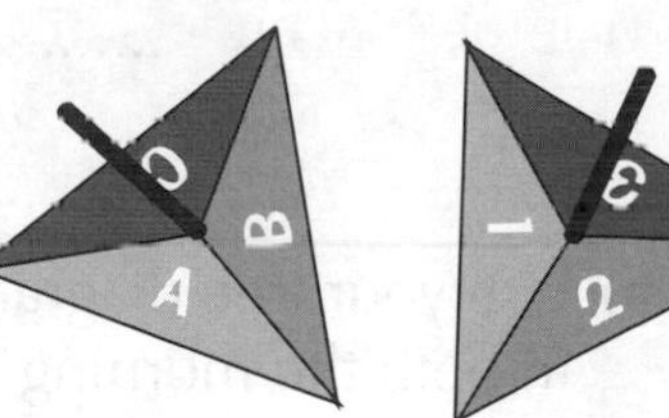

Tree Diagrams

Tree diagrams are useful when there's more than one event going on. Just like in these questions...

Q1 Michael has five cards, numbered 1 to 5. He picks one card at random, replaces that card back in with the others, and then picks a second card. The tree diagram below shows the probability of Michael getting an even or an odd number with each pick.

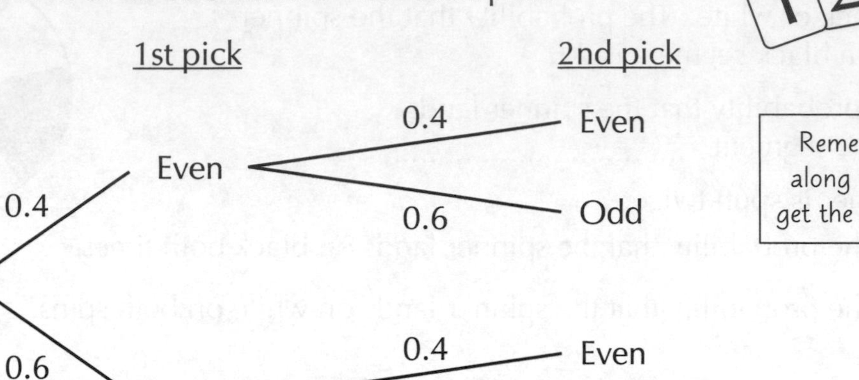

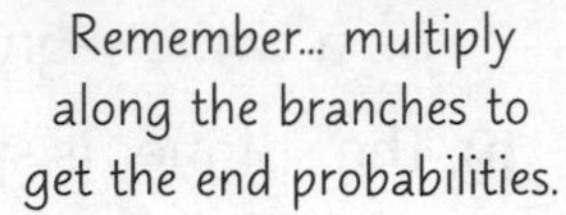

Use the tree diagram to find the probability that Michael:

a) gets an odd number with both picks. ....

b) gets an even number with both picks. ....

c) gets an odd number on the first pick and an even number on the second. ....

d) an odd number and an even number, in either order. ....

Q2 Ania travels to and from work by bus.

- The probability that the bus is late in the morning is always 0.2.
- If the bus is late in the morning, the probability that the bus is late in the evening is 0.7.
- If the bus is not late in the morning, the probability that the bus is late in the evening is 0.1.

a) Fill in the missing probabilities on the tree diagram below.

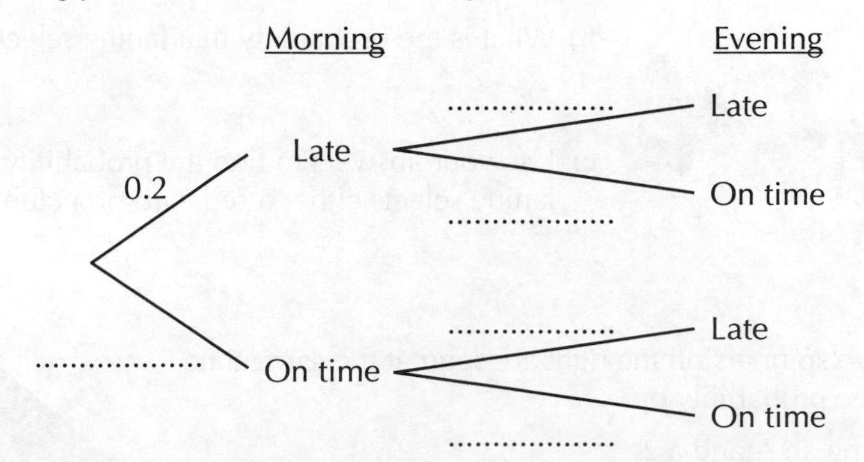

b) Use your tree diagram to find the probability of the bus being on time in both the morning and the evening.

....

Sets and Venn Diagrams

Venn diagrams use overlapping circles to show what things are in what sets.

Q1 A survey asked people if they like bananas and pies.
- 7 people said they like both
- 12 people said they like bananas
- 11 people said they like pies
- 3 people said they like neither

Complete the Venn diagram below to show this data.

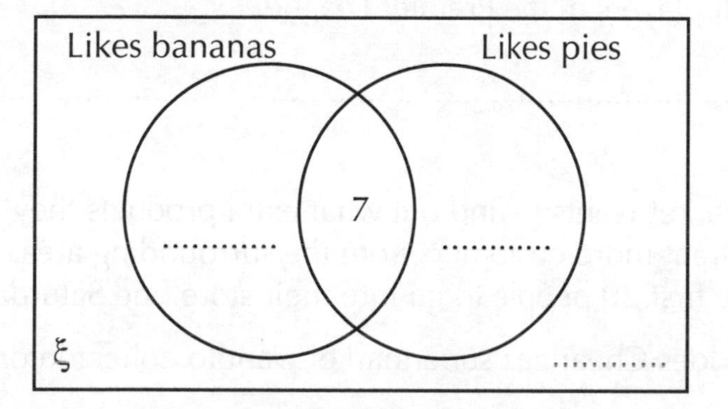

Q2 The whole numbers 1 to 9 are to be put into the following sets:
- A = {numbers that are odd}
- B = {numbers that are multiples of 3}

a) Label this Venn diagram to show how many numbers will be in each of its areas.

Always put the value in the 'overlapping' part of the diagram first.

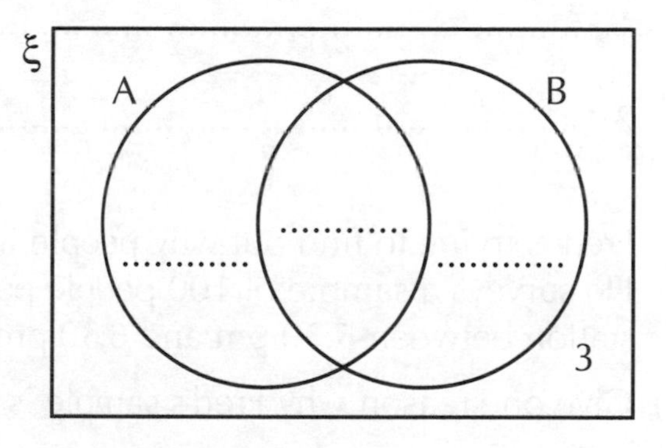

b) Alexei has a set of cards numbered 1-9.
He chooses one card at random. P(A) is the probability of picking an odd number and P(B) is the probability of picking a multiple of 3. Use your Venn diagram to find:

i) P(A) ....

ii) $P(A \cap B)$ ....

iii) P(B′) ....

$A \cap B$ is the intersection of sets A and B. So $P(A \cap B)$ is the probability that a card is an odd number and a multiple of 3.

Q3 All 100 students in a year group sat a test in English and a test in Science.
- 39 of the students passed both tests.
- 75 of the students passed the English test.
- 57 of the students passed the Science test.

a) Show this information on a Venn diagram.

b) One student is chosen at random from the year group.
Find the probability that she only passed one of the tests. ....

Sampling and Bias

Sampling means using information about small groups to tell you about big groups. It's great... but you have to choose your small groups carefully.

Q1 Say what the population is for each of these surveys.

a) The health effects of smoking on 20- to 30-year-old women in the UK.

....

b) The pay of football players in the Premier League.

....

Q2 Cheapeez supermarket wants to find out what extra products they should stock to attract more customers from the surrounding area. They interview the first 20 people to go into their store one Saturday morning.

a) What population does Cheapeez supermarket want to collect information about?

....

b) Give two reasons why their sample might produce biased data.

1. ....

2. ....

Q3 Fred is trying to find out why people in a town use public transport. He surveys a sample of 100 people passing through the town bus station between 5:30 pm and 6:30 pm on a Monday evening.

a) Give one reason why Fred's sample is biased.

....

b) Say how Fred could improve his sample to avoid bias.

....

Q4 A simple random sample of 20 students from a school is needed. The sample is to be selected by:

- Assigning a different number to every student in the school.
- Creating a list of 20 random numbers.
- Selecting the students whose numbers matched the random numbers.

The 20 random numbers can be generated by picking numbers out of a bag. Give one other way of generating random numbers.

....

Collecting Data

You need to record data in a way that makes it easy to analyse.

Q1 Zac collects some data about his school. The data items are listed below.
Tick the correct box to say whether each data item is qualitative or quantitative.

a) The colours of pants worn by the teachers. ☐ **Qualitative** ☐ **Quantitative**

b) The number of students late to school on the first day of term. ☐ **Qualitative** ☐ **Quantitative**

c) The distance travelled to school by each student. ☐ **Qualitative** ☐ **Quantitative**

Q2 Amy collects some data at her school sports day. The data items are listed below.
Tick the correct box to say whether each data item is discrete or continuous.

a) The number of competitors in each event. ☐ **Discrete** ☐ **Continuous**

b) The distance thrown by each competitor in the javelin. ☐ **Discrete** ☐ **Continuous**

Q3 Fred asked each of his 20 classmates how long (in minutes) it took them to eat their dinner.

Here are the results he recorded:

42	13	6	31	15
20	19	5	50	14
8	25	16	27	4
45	32	31	31	10

Group the data appropriately and fill in the table.

Length of time (mins)	Tally	Frequency

Q4 Stanley is researching the use of the local leisure centre.
He asks two questions to a sample of people in his street.
Give one criticism of each of Stanley's questions.

a) ***How often do you use the local leisure centre? Tick one of the boxes.***
Very often ☐ ***Quite often*** ☐ ***Not very often*** ☐ ***Never*** ☐

Criticism: ....

b) ***How old are you?***
☐ ***Under 18*** ☐ ***18 to 30*** ☐ ***30 to 40*** ☐ ***40 to 60*** ☐ ***over 60***

Criticism: ....

Mean, Median, Mode and Range

This page and the next are all about the different types of average — the mean, the median and the mode. And as a free bonus gift, there are questions about the range too.

Q1 Find the mode and the range for each of these sets of data.

a) 3, 5, 8, 6, 3, 7, 3, 5, 3, 9

Mode = ... Range = ...

b) 52, 26, 13, 52, 31, 12, 26, 13, 52, 87, 41

Mode = ... Range = ...

Q2 Find the median for these sets of data.

a) 3, 6, 7, 12, 2, 5, 4, 2, 9

.... Median = ...

b) 14, 5, 21, 7, 19, 3, 2, 5

.... Median = ...

Q3 These are the heights of fifteen unicorns.

162 cm 156 cm 174 cm 148 cm 152 cm 139 cm 167 cm 134 cm
157 cm 163 cm 149 cm 134 cm 158 cm 172 cm 146 cm

a) What is the median height?

....

....

Median = ... cm

b) What is the range of the heights?

Range = ... cm

Q4 Find the mean of each of the sets of data below.

Remember the formula for the mean: total of the items ÷ number of items

a) 13, 15, 11, 12, 16, 13, 11, 9

....

b) 16, 13, 2, 15, 5, 9

....

c) 80, 70, 80, 50, 60, 70, 90, 80, 50, 70, 70

....

Mean, Median, Mode and Range

Q5 Dave has been measuring the length of beetles for a science project.
His results are shown in this stem and leaf diagram.

Use the stem and leaf diagram to find:

1	1 2 8 9 9
2	0 2 7
3	1 4

Key: 2 | 2 means 22 mm

a) the range of the data.

....

b) the median of the data.

....

Q6 The cost of ten second-hand cars sold by Honest Autos are shown below.

£800 £700 £900 £1000 £700
£750 £750 £600 £1100 £15 000

a) Find the mean cost of the ten cars.

....

b) Explain why this doesn't give a good idea of the typical cost of a car at Honest Autos.

....

Q7 The length (in minutes) of comedy films and action films were recorded during a study.
Some results of the study are shown below.

<u>Length of comedy films</u>
Mean = 96
Median = 98
Range = 32

<u>Length of action films</u>
Mean = 124
Median = 118
Range = 54

Compare the two distributions.

....

....

Q8 Colin's mean mark over three exams was 83.
His mean mark for the first two exams was 76.

a) How many marks did Colin score altogether in the three exams?

....

b) How many marks did Colin score altogether in the first two exams?

....

c) What was Colin's score in the final exam?

....

Simple Charts and Graphs

Make sure you read these questions carefully. You don't want to lose easy marks by looking at the wrong bit of the table or chart.

Q1 This pictogram shows the favourite drinks of a group of pupils.

Favourite Drinks	Number of Pupils
Lemonade	◈ ◈ ◈ ◈ ◈ ◈
Cola	◈ ◈ ◈ ◈ ◇
Cherryade	◈ ◈ ◈ ◈ ◈ ◇
Orange Squash	◈ ◈ ◈
Milk	

Don't forget to use the pictogram's key.

◈ Represents 4 pupils

a) How many pupils' favourite drink was lemonade? .... pupils.

b) Milk was the favourite drink of 6 pupils. Show this on the pictogram.

c) How many pupils were asked in total? .... pupils.

Q2 Megan records whether people choose pizza or pasta in her Italian restaurant. The dual bar chart shows the results over one week.

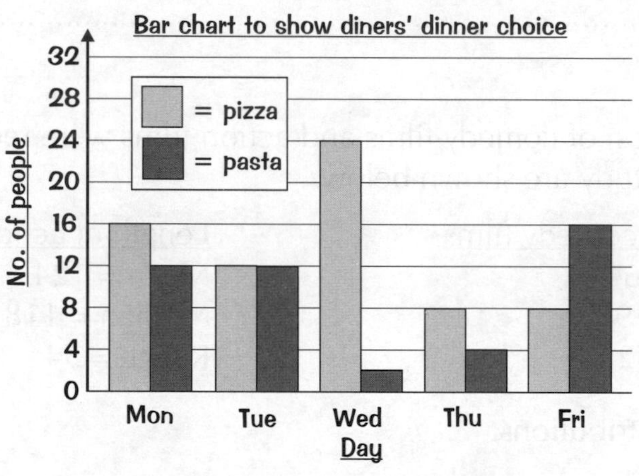

a) On which days was pizza more popular? ....

b) On which day did Megan sell the most pizza and pasta overall? ....

Q3 The graphs below show the marital status of men and women aged 65 and over.

a) What percentage of males aged 65-69 are married?

....

b) What percentage of females aged 75-79 are divorced or separated?

....

Males / Females

England and Wales Percentages: 0, 20, 40, 60, 80, 100

65-69, 70-74, 75-79, 80-84, 85-89, 90 and over

Never married, Widowed, Divorced/separated, Remarried, Married

Simple Charts and Graphs

Q4 The local library carried out a survey to find the types of books borrowed one day. Complete the bar chart on the right showing these results.

Category	Number of books
Crime	10
Fantasy	8
History	3
Romance	4
Science	7

Number of books: 2, 4, 6, 8, 10

Category: Crime, Fantasy, History, Romance, Science

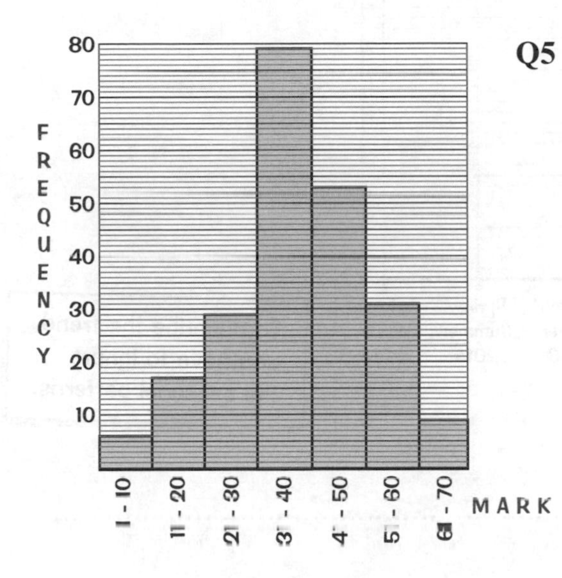

Q5 The bar chart on the left shows the marks scored by some students in an English test.

a) How many students scored 20 marks or less? ....

b) The pass mark for this test was 31. How many students passed the test? ....

c) How many students took the test? ....

d) What is the probability that a randomly selected student's mark was in the modal class? ....

Q6 One hundred vehicles on a road were recorded as part of a traffic study. Complete this two-way table and use it to answer these questions.

	Van	Motor-bike	Car	Total
Travelling North	15			48
Travelling South	20		23	
Total		21		100

a) How many vans were recorded? ....

b) How many vehicles in the survey were travelling south? ....

c) How many motorbikes were travelling south? ....

d) How many cars were travelling north? ....

Q7 A number of students were asked how many times they visited the school cafeteria in a week. The results are shown in this line graph.

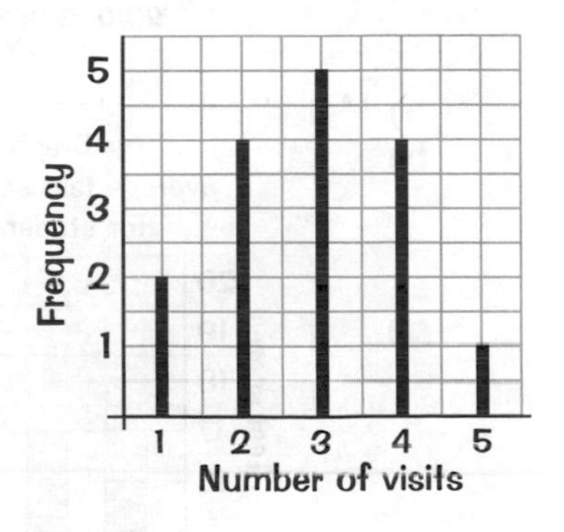

a) What was the modal number of visits to the cafeteria? ....

b) How many students made 2 visits to the cafeteria? ....

c) How many visits were made altogether? ....

d) Calculate the mean number of visits made. ....

e) What is the probability that a randomly selected student visited the cafeteria the median number of times? ....

Simple Charts and Graphs

Q8 The table below shows the amount spent on sun cream over four years by a British actor living in Hollywood.

Season	Spring/ Summer 2011	Autumn/ Winter 2011	Spring/ Summer 2012	Autumn/ Winter 2012	Spring/ Summer 2013	Autumn/ Winter 2013	Spring/ Summer 2014	Autumn/ Winter 2014
Amount spent ($)	33	18	28	17	25	12	21	11

a) Draw a line graph of the data on these axes to make a time-series plot.

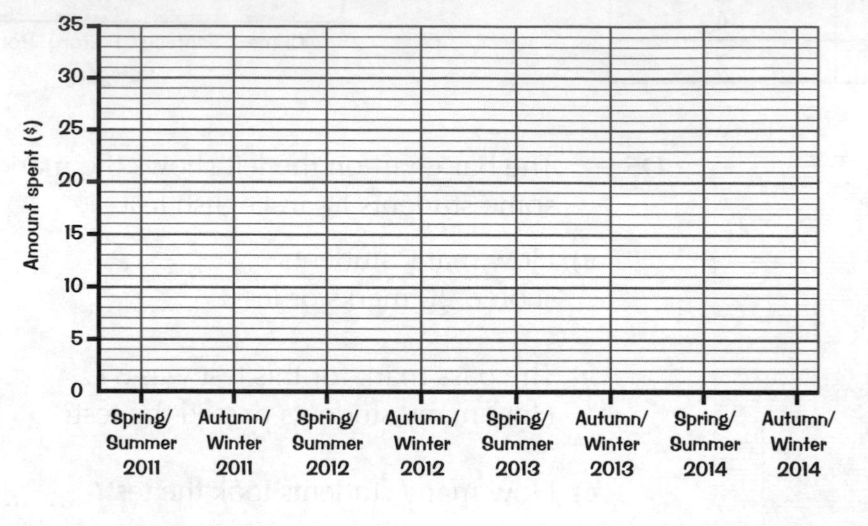

To describe the trend, you have to ignore the seasonal patterns.

b) Describe the trend in the data.

....

....

Q9 Give one reason why each of the following graphs could be misleading.

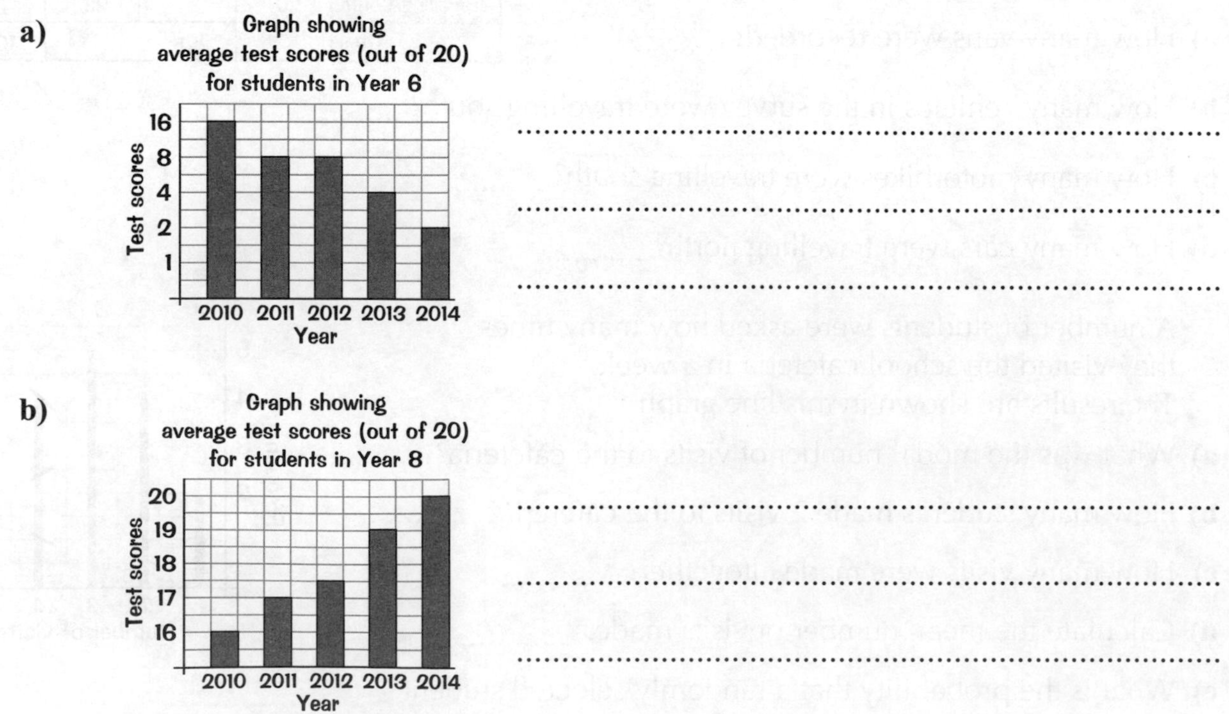

Pie Charts

In a pie chart, the total of everything = 360°. Remember that and you won't go far wrong.

Q1 Sandra is giving a presentation on her company's budget. She has decided to present the budget as a pie chart. The company spends £54 000 each week on various items, which are shown on this pie chart.

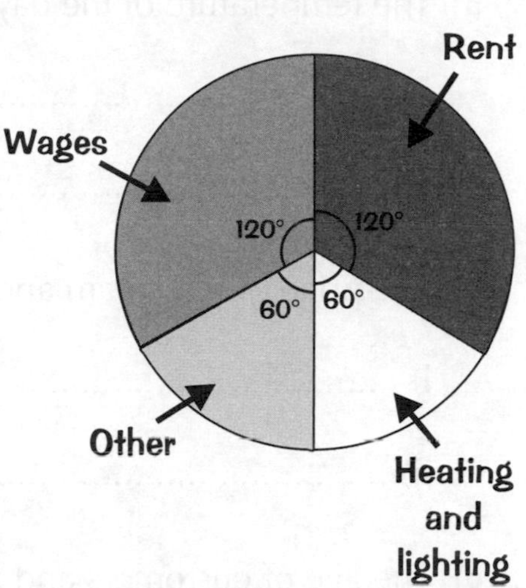

a) What fraction of the budget is spent on wages each week? Give your answer in its simplest form. ....

b) How much money is spent on wages each week? ....

c) What fraction of the budget is spent on heating and lighting each week? Give your fraction in its simplest form. ....

d) How much money is spent on heating and lighting each week? ....

Q2 Pupils at a school were asked about their activities at weekends. The results are shown in the table.

Complete the table and then draw the pie chart using a protractor.

ACTIVITY	HOURS	WORKING	ANGLE
Homework	6	(6 ÷ 48) × 360° =	45°
Sport	2		
TV	10		
Computer games	2		
Sleeping	18		
Listening to music	2		
Paid work	8		
Total	48		

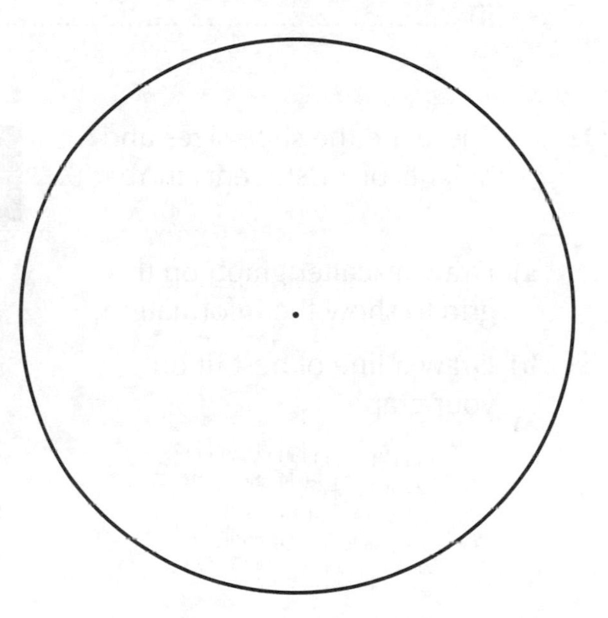

Q3 These pie charts appear in a newspaper article about a local election.

Nicki says that more people voted for the Green party in 2010 than in 2005.

Comment on whether it's possible to tell this from the pie charts.

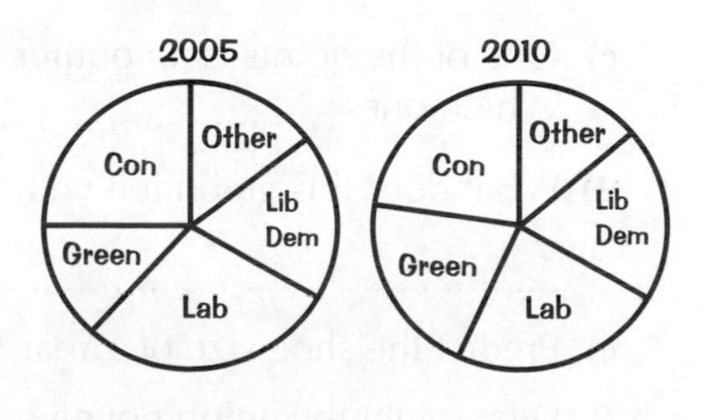

....

....

Scatter Graphs

Q1 For each of the scatter graphs below:
i) describe the correlation shown in each graph,
ii) say what each graph tells you.

Correlation means 'how two things are related'.

a) The temperature of the day and the amount of ice cream sold.

i) ....

ii) ....

b) The price of ice cream and the amount sold.

i) ....

ii) ....

c) The age of customers and the amount of ice cream they buy.

i) ....

ii) ....

Q2 These are the shoe sizes and heights of 11 students in Year 9.

Shoe size	5	6	6	7	7	8	3	5	9	10	10
Height (cm)	155	157	159	158	175	162	149	152	165	167	172

a) Draw a scatter graph on this grid to show the information.

b) Draw a line of best fit on your graph.

A line of best fit should be close to as many points as possible (except outliers).

c) One of the points is an outlier. Which one? ....

d) What does this graph tell you about the relationship between shoe size and height?

....

e) Predict the shoe size of a Year 9 student who's 150 cm tall. ....

f) Why might the graph not give a reliable prediction for the height of a Year 9 student who takes size 12 shoes?

....

Frequency Tables — Finding Averages

Remember... frequency is just another way of saying 'how often something happened'. So a frequency table is just a table telling you how often various things happened. Easy.

Q1 The frequency table below shows the number of hours spent Christmas shopping by 100 people surveyed in a town centre.

Number of Hours	0	1	2	3	4	5	6	7	8
Frequency	1	9	10	10	11	27	9	15	8
Hours × Frequency									

a) What is the modal number of hours spent Christmas shopping? ....

b) Fill in the third row of the table.

c) What is the total amount of time spent Christmas shopping by all the people surveyed? ....

d) What is the mean amount of time spent Christmas shopping by a person?

....

Q2 Last season Newcaster City played 32 matches. The number of goals they scored in each match is shown in this table.

GOALS	FREQUENCY
0	7
1	11
2	6
3	4
4	3
5	1

a) What was the modal number of goals scored in each match? ....

b) What was the median number of goals scored in each match? ....

c) What was the range of goals scored over the whole season? ....

d) i) Add an extra column to the table showing 'Number of goals × Frequency'.

ii) Use this extra column to calculate the total number of goals scored during the season.

....

iii) Calculate the mean number of goals scored per game.

....

Grouped Frequency Tables

Take your time with these questions — grouped frequency tables can easily catch you out.

Q1 Dean is carrying out a survey for his geography coursework. He asked 80 people how many miles they drove their car last year to the nearest thousand miles. He has started filling in a grouped frequency table to show his results.

No. of Miles (thousands)	1 - 10	11 - 20	21 - 30	31 - 40	41 - 50	51 - 60	61 - 70	71 - 80	81 - 90	91 - 100
No. of Cars	2	3	5	19	16	14	10	7		

a) Complete Dean's table using the following information:
81 000, 83 000, 90 000, 91 000

You'll need to <u>divide</u> these numbers by 1000 to figure out which groups they lie in.

b) Write down the modal class. ....

c) Which class contains the median number of miles driven? ....

d) Dean claims that 90% of the people in his survey drove more than 30 000 miles last year. Does the data support Dean's claim?

....

Q2 The times (t, in minutes) taken to complete a test by 23 pupils at Blugdon High are shown in this table.

a) Add two extra columns to the table.

i) In the first extra column, write down the mid-interval values for each of the classes.

ii) In the second extra column, multiply the frequency by the mid-interval value for each of the classes.

Time (t)	Frequency
$30 \le t < 40$	4
$40 \le t < 50$	7
$50 \le t < 60$	8
$60 \le t < 70$	4

b) Use the table to estimate the mean time to 1 d.p.

....

c) Estimate the range of times. ....

Q3 This table shows times for two teams of swimmers, the Dolphins and the Sharks.

Dolphins				Sharks			
Time interval (seconds)	Frequency (f)	Mid-interval value		Time interval (seconds)	Frequency (f)	Mid-interval value	
$14 \le t < 20$	3	17		$14 \le t < 20$	6	17	
$20 \le t < 26$	7	23		$20 \le t < 26$	15	23	
$26 \le t < 32$	15			$26 \le t < 32$	33		
$32 \le t < 38$	32			$32 \le t < 38$	59		
$38 \le t < 44$	45			$38 \le t < 44$	20		
$44 \le t < 50$	30			$44 \le t < 50$	8		
$50 \le t < 56$	5			$50 \le t < 56$	2		

a) Complete the table, writing in all mid-interval values.

b) Use the mid-interval technique to estimate the mean time for each team to 1 d.p.

....

MCFW46_MQFW46_MXFW46